RECENT PROGRESS IN ORGANOMETALLIC CHEMISTRY

Edited by **Mohammed Muzibur Rahman**
and **Abdullah Mohamed Asiri**

Recent Progress in Organometallic Chemistry

http://dx.doi.org/10.5772/65606
Edited by Mohammed Muzibur Rahman and Abdullah Mohamed Asiri

Contributors

Vernon Somerset, Ruby Srivastava, Melinda Kovacs, Emoke Dalma Kovacs, Betül Çalişkan, Ming-Der Su, Jia-Syun Lu, Ming-Chung Yang, Shih-Hao Su, Xiang-Ting Wen, Jia-Zhen Xie, Tomasz Goslinski, Wojciech Szczolko, Tomasz Koczorowski

Published by InTech
Janeza Trdine 9, 51000 Rijeka, Croatia

Publishing Process Manager Nina Kalinic
Technical Editor SPi Global
Cover InTech Design team

First published July, 2017
Printed in Croatia
Legal deposit, Croatia: National and University Library in Zagreb

Additional hard copies can be obtained from orders@intechopen.com

Recent Progress in Organometallic Chemistry, Edited by Mohammed Muzibur Rahman and Abdullah Mohamed Asiri
p. cm.
Print ISBN 978-953-51-3317-9
Online ISBN 978-953-51-3318-6

For EU product safety concerns:
IN TECH d.o.o., Prolaz Marije Krucifikse Kozulić 3, 51000 Rijeka, Croatia,
info@intechopen.com or visit our website at intechopen.com.

Meet the editors

Mohammed Muzibur Rahman received his Ph.D. from the Chonbuk National University, South Korea, in 2007. After his Ph.D., he worked as postdoctoral fellowship and assistant professor in pioneer research centers and universities located in South Korea, Japan, and Saudi Arabia (2007 to 2011). Presently, he is working as associate professor in Center of Excellence for Advanced Materials Research (CEAMR) and Chemistry department at King Abdulaziz University, Saudi Arabia, since 2011. He published more than 180 research articles and several proceeding in well-known high-impact ISI-journals; attended more than 60 international and domestic conferences; several patents, and published several book chapters and seven-books as an editor. His research work has been largely in the area of Organometallic chemistry, Electrochemistry, Nanotechnology, Sensors, Surface Chemistry, Instrumental Science, Nanomaterials, Self-Assembled monolayers, Carbon nanotubes, Colloids, Photochemistry, m-chips, and devices etc.

Abdullah Mohamed Asiri received PhD from University of Wales, College of Cardiff, UK in 1995. He is the Head of the Chemistry Department at King Abdul Aziz University since October 2009 and he is the founder and the Director of the Center of Excellence for Advanced Materials Research (CEAMR). He is a Professor of Organic Photochemistry. His research interest covers color chemistry, synthesis of novel photochromic, thermochromic systems, synthesis of novel colouring matters and dyeing of textiles, materials chemistry, nanochemistry, nanotechnology, polymers, and plastics. He is the Editor-in-Chief of King Abdul Aziz University Journal of Science. He is also a member of the Editorial Board of Pigments and Resin Technology (UK), Organic Chemistry in Sight (New Zealand), Recent Patents on Materials Science (USA). He is the Vice-President of Saudi Chemical Society (Western Province Branch).

Contents

Preface

It gives us immense pleasure in introducing a book named *Recent Progress in Organometallic Chemistry* based on the state of the art of organometallic and state-of-the-art catalysis with various inorganic catalyst as well as their numerous potential and significant applications. These deal with aspects of the synthesis, characterization, and applications. The discussion of these aspects develops through the fundamentals and applied experimental routes in catalysis approaches via the synthetic matrixes to include the interfacing of the organic-inorganic chemistry world. Usually, catalysis has certainly attained accomplishment of its classical essence and has taken new directions of synthesis, growth, development, continuous changes, and various catalytical applications in research and development sectors. The new route and emerging frontiers are branching out from time to time around this innovative organometallic, biocatalytic, electro-catalysis technology based on various organic-inorganic matrixes. This work aims to bridge the gap between graduate and researchers in applied catalysis (electro-, photo-, or nanocatalyst) branches as well as medical and metallocenes, in order to initiate researchers into catalysis study in as straightforward way as possible and as to introduce the scientist to the opportunities offered by the applied industrial and technological fields. We worked consistently to complete this work on *Recent Progress in Organometallic Chemistry* under the InTech Open Access Publisher. This book offers a comprehensive but easy-to-follow theoretical context with practical investigation in catalysis science. It frameworks the significant research methods and contemplates technological implementations in research or industrial scales. A straightforward acquaintance of the principles of chemical, physical, biological, and environmental catalysis designed with inorganic chemistry is anticipated. It will also entreat to a broad audience of undergraduate, graduate, and postgraduate students at colleges and universities of technology as well as scientists in industry who prerequisite a comprehensive contingent study in the catalysis science and technology. We hope that the book on *Recent Progress in Organometallic Chemistry* would further enhance the applied metallocenes or nanocatalyst or catalysts in chemical, biological, and medical sciences, especially in bringing new entrants into the applied catalytic science and technology fields, and help scientists to forward and develop their own field of specialization.

Prof. Mohammed Muzibur Rahman, PhD and **Prof. Abdullah Mohamed Asiri, PhD**
Center of Excellence for Advanced Materials Research (CEAMR) and Chemistry Department,
Faculty of Science, King Abdulaziz University,
Jeddah, Saudi Arabia

Section 1

State-of-the-Art: Organometallics

Chapter 1

Theoretical Insight into the Medicinal World of Organometallics: Macro versus Nano

Ruby Srivastava

Additional information is available at the end of the chapter

http://dx.doi.org/10.5772/67781

Abstract

Due to the unique physicochemical properties, organometallic complexes have been widely used in the medicinal world. These complexes have specific properties such as structural diversity, redox/catalytic activities, and possibility of ligand exchange. As the cancer therapies provided by these complexes are not always effective and have desired side effects, new treatment methods are needed for the successful therapies. Recent advances suggest that nanotechnology has also profound impact on the disease prevention, diagnosis, and treatment. The delivery system based on nanotechnology has faster drug absorption, controlled dosage release, and minimal side-effects. This technology is used for the treatment of cancer till now, but soon, it will find applications to other diseases also. The use of nanotechnology in the field of drug delivery is to develop a system that improves the solubility and bioavailability of hydrophobic drugs. It is used to increase specificity, developing delivery system for slow release, and to design delivery vehicles that can improve the circulatory presence of drugs. As the photophysics of organometallic complexes is still not clear, this topic is included to discuss the latest developments in this field, which allows the photochemical reactions at the nanolevel.

Keywords: organometallic complexes, nanotechnology, photophysics, drug delivery, upconversion luminescence

1. Introduction

Organometallic chemistry deals with three basic aspects as environmental concern, biological aspect, and medicinal chemistry. Medicinal organometallic chemistry continues to be a major application for these compounds in biology. Medicinal organic chemistry has therapeutics, diagnostics, and theranostics effects. Medicinal organometallic complexes consist of platinum, ruthenium, iron, titanium, and gold among other metals. Fundamental studies have been carried out on the organometallic complexes in which the mechanism of action exert their medicinal effect (e.g., induce cell death in cancer cells), the synthesis of new organometallic compounds and the development of combination therapies containing organometallic components. Research has shown significant progress in utilization of transition metal complexes as

1.1. Anticancer agents

The development of metal complexes with platinum as a central atom such as cisplatin or carboplatin had an enormous impact on current cancer chemotherapy. Cisplatin has become one of the most widely used drugs and is highly effective in treating several cancers such as ovarian and testicular cancers. The limitations of cisplatin have stimulated research in the field of platinum antitumor chemistry by including the reduction in toxicity of cisplatin (nausea, ear damage, vomiting, loss of sensation in hands, and kidney toxicity), acquired drug resistance observed in certain tumors and inefficiency of the drug against some of the commonest tumors (e.g., colon and breast). Due to its particular chemical structure, cisplatin offers little possibility for improvement in tumor specificity and thereby reducing side effects. The other alternative complexes have at least one direct, covalent metal-carbon bond, having structural variety, diverse stereochemistry, provide control over major kinetic properties, kinetically stable, usually uncharged, and having low oxidation state of metal atom, so they can be used as ideal candidates for anticancer candidates [1–4]. Some examples are: metallocenes [5–9], organometallic ruthenium half-sandwich complexes [10–14], organometallic osmium half-sandwich complexes [15–17], organometallic iridium and rhodium complexes [18–21], rhenium organometallics [22–24], ruthenium, osmium, iridium, and platinum organometallics as scaffolds for protein kinase inhibitors, metal NHC complexes [25, 26], and metal carbonyl complexes [27, 28].

1.2. Antibacterial agents

The biggest challenge in the antibacterial market is the issues related to the drug-resistant pathogens. The remedy is now to search for new compounds with new mode of action to overcome the resistant strains. This can be done by either the organic derivatization of old drugs or completely new organometallic drugs, for example, new tamoxifen [19–21, 29], platensimycin [22–24], etc.

1.3. Anti-infectant agents

Transition metal as silver is being used as the antimicrobial agent due to its low toxicity as compared to the other metals. For example: silver (1) sulfazine, which is used to treat burns to prevent the bacterial infections. Silver nitrate is given to the infants to prevent the development of ophthalmia neonatorum. Chlorhexidine-silver sulfadiazine is an anti-infective metal complex against catheter infections. Organometallic complexes of Pt [30–32], Rh, Ir, Pd, and Os metal with active organic molecules have been reported to exhibit trypanocidal activity. Metal complexes of Pt (II) and Ru (II) with o-vanillin-(4-methyl thiosemicarbazone), and o-vinillin-(4-phenyl thiosemicarbazone), metal complexes of Ga (III), Al (III), and Fe are among the various other drugs [33–36].

1.4. Anti-inflammatory agents

These complexes are also used as anti-inflammatory and antiarthritic agents. Several injectable transition metal complexes as sodium aurothiomalate, aurothioglucose, sodium aurothiopropanol and gold and silver nanoparticles conjugated with heparin derivative possess antiangiogenesis properties [37–39]. Gold has been used for the treatment of peripheral psoriatic arthropathy. As a product of oxygen metabolism, superoxide anion can trigger oxidative injury to tissues. This activity is associated with riper fusion and inflammatory diseases as well as neurological disorders such as Parkinson's disease and Alzheimer's disease. However, excess use of these complexes in arthritis causes pain and fever. NO is excellent ligand for transition metal ions and these metal nitrosyls having therapeutic values. Sodium nitroprusside is used to treat cardiovascular diseases by releasing NO with limited usage due to the toxicity of CN^-. Ruthenium poly aminocarboxylate complexes are also efficient NO scavengers [40, 41].

1.5. Anti-diabetic agents

Diabetes is the most suffering disease among human beings. This is the disease in which the body do not produce insulin hormone, which is used for absorption of glucose in cells. The control of glucose level is done by vanadium complexes with organic ligands which are less toxic and have improved solubility and lipophilicity [42]. These complexes show involvement in the activation of prominent key components of insulin-signaling pathways [43]. Chromium supplementation also improves glycemia among patients of diabetes [44]. Similarly, higher zinc intake also lowers the risk of type 2 diabetes in women [45].

1.6. Neurological drugs

Neurological disorders are also treated by transition metal complexes. Lithium is used for Huntington's chorea, tardive dyskinesia, spasmodic torticollis, Tourette's syndrome, L-dopa induced hyperkinesia, Parkinsonism, organic brain disorders, drug induced delusional disorders, migraine and cluster headache, periodic hypersomnolence, epilepsy, Meniere's disease, and periodic hypokalemic paralysis. Lithinium inhibit the scavenging pathways for capturing inositol in the resynthesis of polyphosphoinositides in the brain. Zinc is also used as transmitter in neuronal signaling pathways.

1.7. Delivery probes and diagnostic tools

Organometallic complexes have unique properties as redox activity, Lewis acidity, electrophilicity, valency, geometry magnetic spectroscopic, and radiochemical properties which can be used to measure cellular functions. Gold nanorods has been used for photoacoustic molecular imaging with simultaneous multiple targeting as they are less reactive and less toxic. The nanoparticles injected in the tumor cells increases their ability to absorb radiation of specific wavelength. The property is used in lymphotropic nanoparticle-enhanced magnetic resonance imaging of prostate cancer. As iron oxide has superparamagnetic properties so they can act as negative contrast agent in magnetic resonance imaging (MRI) which is used to detect the sensitivity of inflamed tissues.

Transition metals exhibit different oxidation states and can interact with several negatively charged molecules. Due to their vital role in medicinal chemistry, we have included both the macro and the nanoorganometallic complexes and their structural and photophysical behavior in detail.

Metallocene compounds have two π-bonded cyclopentadienyl (Cpa) ligands on a metal atom. These compounds are also called "sandwich complexes" due to their symmetrical nature. Other metal complexes with cyclic π-perimeters are also named as metallocenes. Compounds with only one π-perimeter are classified as "half sandwich metallocenes." The bis-cyclopentadienyl complexes are divided in two categories: (a) "classical" with parallel Cp rings and (b) "bent" metallocenes, which have other ligands bonded to the metal in addition to the Cp rings. Ferrocene was the first organometallic compound with antiproliferative properties, so the medicinal properties of the complex were investigated [3]. Ferrocene is nontoxic compound and can be injected, inhaled, or taken orally. It cannot cause major health problems [4, 5]. Another ferrocene-containing compound chloroquine (derivative) is used as antimalarial drug. Ferroquine has an activity-like chloroquine on the malaria parasite *P. falciparum*. The p-methoxybenzyl substituted titanocene show very good activity against renal cell cancer and pleura mesothelioma cell lines. Ruthenium complexes have low toxicity and it has the same mechanism (ligand exchange kinetics) to those of platinum(II) antitumor drugs [10]. A class of ruthenium(II)-arene complexes that are weakly cytotoxic *in vitro*, [Ru(η6-p-arene)Cl_2(1,3,5-triaza-7-phosphaadamantane)] termed as RAPTA, interact strongly with proteins, with the ability to discriminate binding to different proteins, but show a relatively low propensity to bind DNA, which is considered to be the main target of many metal-based drugs. Dyson et al. recently described the preparation of a series of RAPTA-type complexes with fluoro-substituted η6-arene ligands [46] (**Figure 1**). The active Pt-drug seems to be *cis*-coordinated by bidentate amine ligands or two amines (at least one -NH group on the amine) and two leaving groups with an intermediate binding strength (e.g., Cl^-, SO_4^{2-}, citrate or oxalate) to platinum. Nonplatinum metals may have some different chemical behavior (oxidation state, redox potential, coordination geometry, additional coordination sites, binding preferences to biomolecules as the HSAB (hard and soft (Lewis) acids and bases), rate of hydrolysis or kinetics of ligand exchange reactions and the ability to replace essential metals. Due to these differences, the nonplatinum metal-based compounds may have different mechanisms of action, biodistribution, and biological activity.

Studies were carried out on complexes with iron, cobalt or gold, titanium, ruthenium, or gallium central atoms, which have shown the promising results in preclinical studies. Other metal complexes which have shown potential anticancer activity are the complexes of Rh (I), Rh (III) [22, 23, 47], Ir (I), Ir (II), Ir (IV) [48, 20, 21], Os (II), and Os(III) [18, 49–52]. Ferrocifenes [53] exhibit anticancer activity against hormone dependent and hormone-independent breast cancers. Ferrocene derivatives as curcuminoids [54], androgen derivatives [55], and antiandrogens derived from the nilutamide lead structure [56],

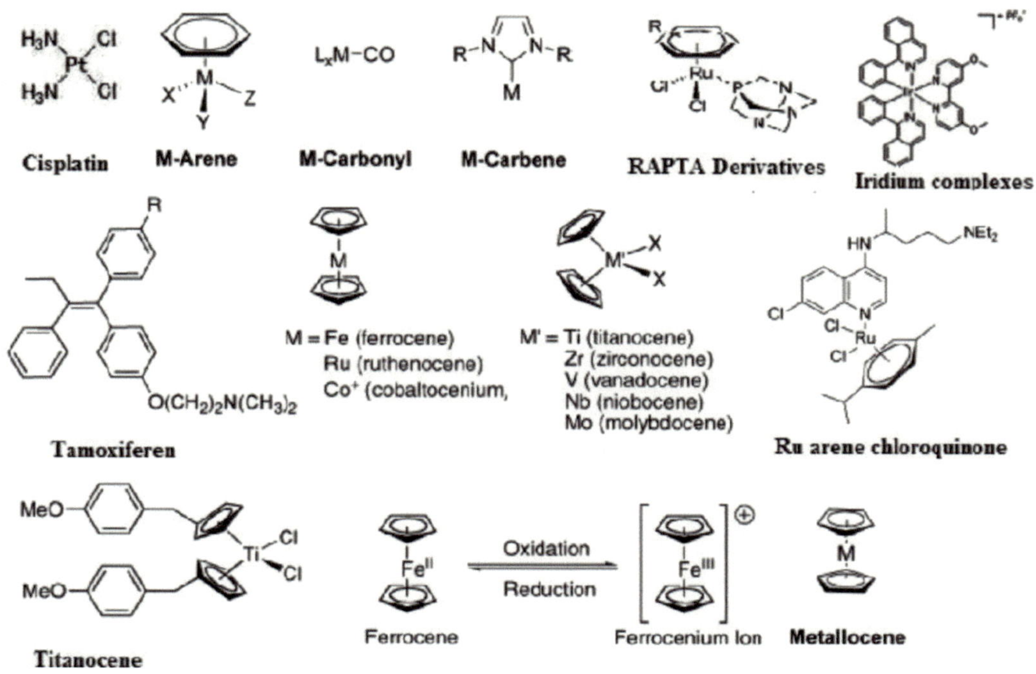

Figure 1. Schematic diagram of few organometallic complexes.

indolones [57], and ferrocenophane polyphenols [58] has also been used for antiproliferative activities.

Transition metal carbene complexes also feature a divalent organic ligand, which is coordinated to the metal center. As these complexes are highly stable and easily derivatize, they can be the suitable candidates for drug development [8, 59].

Metal NHC complexes are also having pharmacological properties as novel antibacterial and antitumor drugs. Their mode of action is both coordinated metal–respective biological target-dependent thioredoxin reductase or other enzymes containing (seleno) cysteine residues in their active site for gold or DNA for copper NHC complexes (half-sandwich) [60, 32], ruthenium [27], or manganese [28, 61] bioorganometallic species and complex containing an acetylsalicylic acid (aspirin) derived ligand emerged as cytotoxic drugs.

An enormous work has been carried out by my mentor Prof. G. Narahari Sastry and group in the field of anticancer treatment. The research group has focused their attention to the biochemical aspects of the clinical application of aromatase inhibitors with designing strategies on toxicity profile, pharmacokinetics, relative potency of aromatase inhibitors, and pharmacophores models [62–73].

As the side effects of these complexes are unavoidable, the research was shifted to the nanotechnology which will have a profound impact on disease prevention, diagnosis, and treatment.

Few advantages of nanotechnology techniques are:

1. Protect drug from degradation
2. Easily changeable physical properties due to nanosizes
3. Reduced dose size
4. Ease of drug targeting due to nanosize
5. Allow delivery of insoluble drugs
6. Longer circulation time
7. Maintain its therapeutic activity
8. Improve the oral bioavailability of the agents
9. Passive targeting of drugs to the macrophages (liver and spleen)

2. Nanoorganometallic complexes

Recent advances suggest that nanotechnology will give a better solution for disease prevention, diagnosis, and treatment. It is an ideal targeting system, should have long circulating time, be present at appropriate concentrations at the target site, and should not lose its activity or therapeutic efficacy while in circulation. The increased vascular permeability coupled with an impaired lymphatic drainage in tumor allows an enhanced permeability and retention effect of the nanosystems in the tumor or inflamed tissue. Nanotechnology offers a solution for using the numerous chemical entities for treating brain disorders that are not clinically useful because of the presence of the blood-brain barrier.

The advantage of nanoparticles with potential MRI-related medical applications comprise of various materials, such as metals (gold, silver, and cobalt) or metal oxides (Fe_3O_4, TiO_2 and SiO_2) (**Figure 2**). Magnetic nanoparticles coated with dimercapto succinic acid (DMSA) were toxic to neurons in a dose-dependent manner. Cobalt (Co), gold (Au@Fe), and platinum (Pt@ Fe) are the other types of nanomaterials that show potential application in antimicrobial and anticancer treatment. Several studies on nanoparticles shown them to be cytotoxic [72], genotoxic [73], and potentially carcinogenic [74] and are used to induce apoptosis and inhibit cell proliferation [75].

These nanomaterials contain the sphere and core-shell structures, two-dimensional (2D) grapheme nanosheets have great potential for high drug loading efficiency and conjugation of proteins, drugs, and fluorescent probes.

The molecular imaging applies to various techniques such as positron emission tomography (PET), computed tomography (CT), or ultrasound and magnetic resonance imaging (MRI) which gives the best spatial resolution and is either noninvasive or minimally invasive. As MRI is not applied in full potential due to low specificity, so it can be alternatively taken to use as cell markers. The unique paramagnetic and superparamagnetic properties of nanoparticles (NP) can be utilized for the detection with MRI in small quantities. Nanoparticles with potential MRI-related medical applications comprise various materials, such as metals (gold,

http://dx.doi.org/10.5772/67781

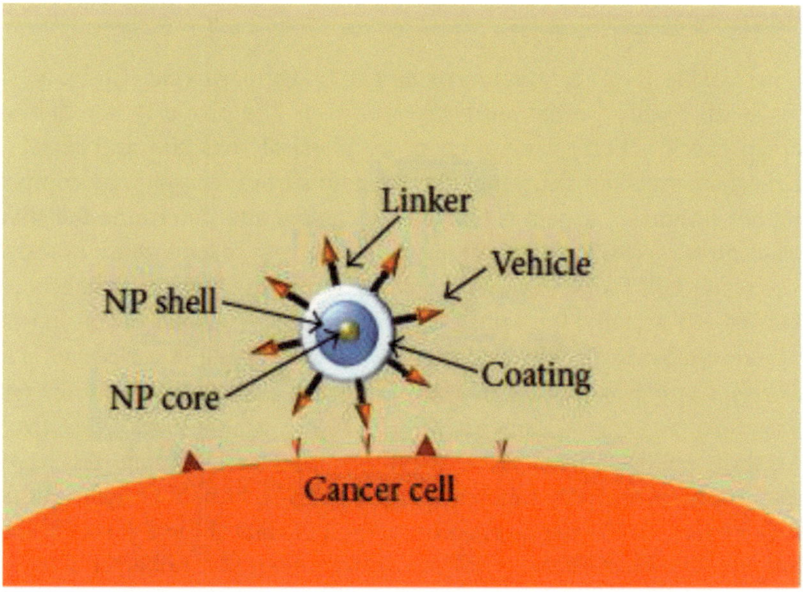

Figure 2. Schematic representation of the targeted contrast agent used for MRI approaching of the cancer cell and specific proteins.

silver, and cobalt) or metal oxides (Fe_3O_4, TiO_2, and SiO_2). While diagnostic is a common medical application of nanoparticles, they can also be used for therapy [76–80] (**Figure 3**).

Nanoparticles can be categorized in two parts:

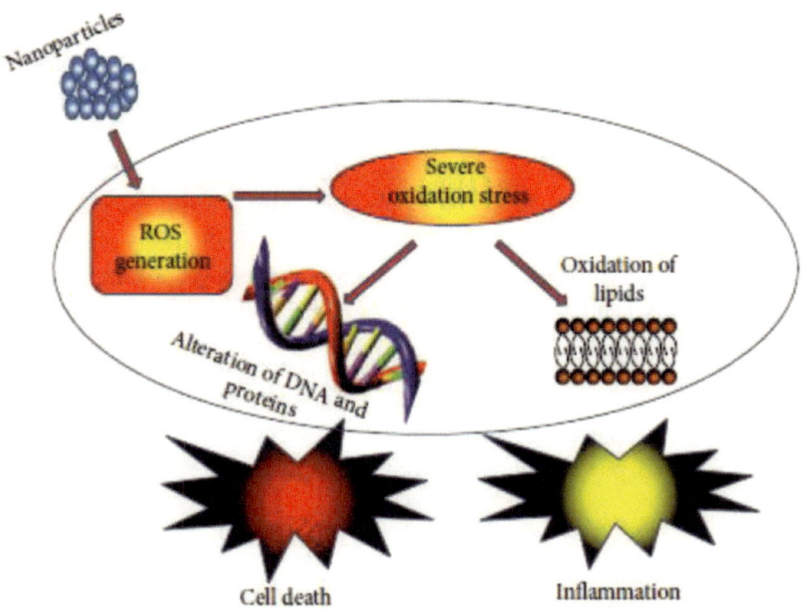

Figure 3. Representation of toxicological mechanisms of NM to eukaryotic cells.

2.1. Inorganic nanoparticles

Silver (Ag), iron oxide (Fe_3O_4), titanium oxide (TiO_2), copper oxide (CuO), and zinc oxide (ZnO) are used for highly potent antibacterial effect. The property is exhibited through reactive oxygen species (ROS) generation or by physical structure and metal-ion release. Though the mechanism is not clear, nonetheless high surface energy may compromise their efficacy. Another important aspect is that how to define and determine the silver minimal inhibitory concentration (MIC) and breaking point, the ease of emergence of resistant strains [81–83]. Silver really kills biofilm or planktonic cells and finally the side effects of silver and its complexes [84–87] remains the same. Yet till now it is the most promising antibacterial nanometal. Titanium oxide (TiO_2) has shown its efficiency against various viral species and parasites [88–90]. Copper oxide (CuO) is less expensive and used for efficacy enhancement [91–93]. Iron oxide (Fe_3O_4) [94], zinc oxide (ZnO) and Magnesium oxide (MgO) [95–97] nanoparticles show antibacterial activities. Gold nanoparticles and nanorods have been used as bactericidal in photothermally functionalized form [98]. Pt nanoparticles diffuse through membranes and induce DNA damage, accumulation of cells at the S-phase of the cell cycle, and apoptosis [99]. The properties of Al_2O_3 are unclear about the antibacterial treatment [100], while SiO_2, Au, Fe_2O_3, and TiO_2 are biocompatible.

Even cytotoxic NM can be converted into biocompatible materials through slight variation in their surface structure. Therefore, we can say that nanomaterials possess a broad level of biological properties that are highly dependent upon their size, structure, quantity, and receptor cell type. Though, the nanomaterials that penetrate the body through the skin by respiration or by inhalation directly affect the major body organs (lungs, heart, and brain).

2.2. Organic nanoparticles

Quaternary ammonium compounds, imidazole derivatives, alkyl pyridiniums, copolymers of N-vinylimidazole and phenacyl methacrylate, benzoic acid, phenol, and p-Hydroxy benzoate esters, quaternary phosphonium or sulfonium groups, triclosan, 5-chloro-8-hydroxy-quinoline, chitosan, or quaternary phosphonium are the polymeric nanoparticles that are used to kill microorganisms either by releasing antibiotics, antimicrobial peptides, and antimicrobial agents or by contact-killing cationic surfaces. Organic antibacterial materials are less stable than inorganic materials at high temperatures [101, 102]. Still some phenomena such as several NM killing pathways, effects of NM's treatment combinations and bacterial intrinsic pathways of programmed cell death in NM's dependent killing are yet to be understood.

As we all know, hepatocellular carcinoma (HCC) is the leading cause of cancer-associated death and the conventional treatment is still not satisfactory due to chemoresistance and recurrence. In a recent study, Pt nanocluster assembly (Pt-NA) composed of assembled Pt nanoclusters was synthesized incorporating a pH-sensitive polymer and HCC-targeting peptide [103] .

The advantage of Pt nanocluster medicine is that Pt-NA is active in peripheral blood and readily targets tumor cells including CLSC because of (i) the surface-targeting peptide; (ii) protonation of pH-sensitive polymers in an acidic intracellular environment triggers Pt-NA disassembly into extremely small Pt nanoclusters; and (iii) the resulting extremely small Pt

http://dx.doi.org/10.5772/67781

nanoclusters with large specific surface accelerate the release of toxic Pt ions inside the cells for an effective cancer treatment (**Figure 4**).

Numerous efforts have been devoted to synthesize nanostructured materials with specific morphology as their size and shape play an important role in determining their functions. It was seen that the cationic nanoparticles with metals (gold, silver, and cobalt) or metal oxides (Fe_3O_4, TiO_2, and SiO_2) were moderately toxic than their anionic nanoparticles. The studies reflect that DMSA coated nanoparticles are nontoxic to HeLa cells or RAW macrophages. The incorporation of chlorotoxin onto functionalized Fe_3O_4 nanoparticles resulted in a significant increase in the total uptake within the brain tumors of mice. Substituted magnetic spinel ferrites of the general formula MFe_2O_4 (where M = Zn^{2+}, Mn^{2+}, Co^{2+}, Ni^{2+}, and Mg^{2+}) offer the opportunity to fine-tune the magnetic properties of the inorganic nanoparticle core as a function of the kind of divalent ion.

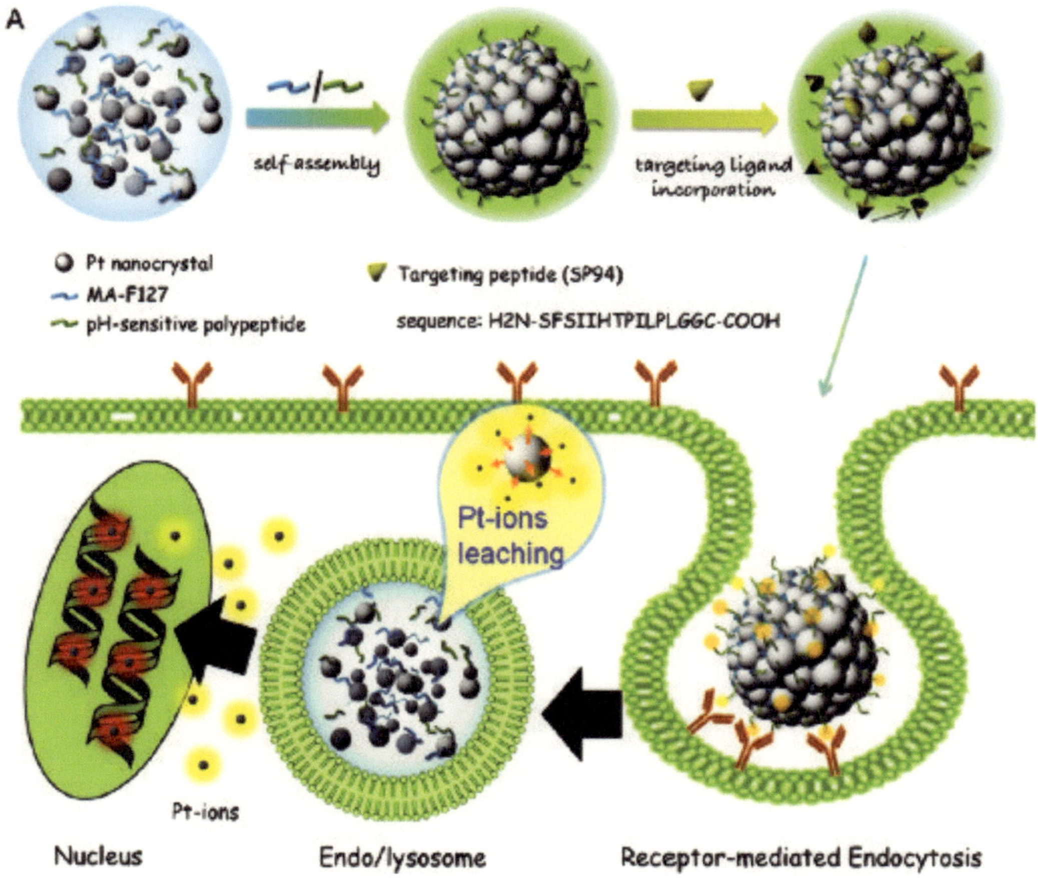

Figure 4. Schematic representation of HCC targeted Pt nanocluster assembly (Pt-NA). Adapted with permission from American Chemical Society [103].

3. Photophysics of the bioactive molecules

In the recent years, new experimental and theoretical developments have occurred in the field of photoactivatable metal complexes which play active role in the field of medicine and biotechnology. Some metal-DNA complexes possess favorable emission properties, while some complexes also provide site-directed therapy. These properties help in oncology, where metal-based precursors generate excited state drugs with different mechanisms.

In this section, the computational techniques (time-dependent density functional theory) and ultrafast-pulsed radiation techniques will be discussed.

The delivery of light depends on the efficiency of light source. It should be efficient to activate the complex. The irradiation should occur in the strong MLCT transitions. However, in medicinal world the UV radiations are harmful but red region is preferred as it deeply penetrates the tissues. Two and three photon absorption can be achieved as the desirable condition is to activate complexes that absorb at shorter wavelength using laser beam that penetrates tissues deeply.

The use of organometallics has become a topic of interest for design of tractable therapeutic agents and theranostics [104]. The most promising organometallic complexes (and motifs) used in cancer therapy is RAPTA-C: [Ru(η6-*p*-cymene)Cl_2(pta)] (pta = 1,3,5-triaza-phosphatricyclo [3.3.1.1]-decane; **Chart 1**), along with its osmium analogue and their corresponding functionalized derivatives [105]. These complexes exhibit antimetastatic properties *in vivo*. The quite strongly bonded phosphine and arene ligands, the chloride ligands rapidly interchange with water molecules. Arene-ruthenium derivatives can react with N- and S-donors so that they can bind to both nucleotides and proteins [106].

Porphyrins and their metalloderivatives are used for photodynamic therapy [107] and optical imaging and as theranostic agents [108]. Gold (III), Palladium (II), Palladium (III) inside the porphyrin rings and their derivatives can act as anticancer agents [109]. It is based on the concept of "optical bi-theranostic" (two modalities for therapy and one for optical imaging). Further as the intramolecular interactions between the two moieties alter their activities so this should be considered for designing and testing. Ruthenium and Iridium possess favorable photophysical properties which allow functional imaging of cells and tissues (e.g., DNA interactions) and provide site-directed therapy. The electronic transitions can be metal-centered (MC), ligand-

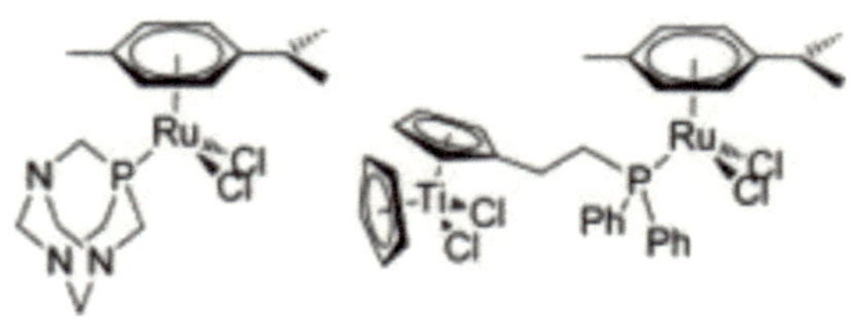

Chart 1. Structure of [Ru(η^6-*p*-cymene) Cl_2(pta)] (left) and [(η^5-Cp)Ti((η^5-C_5H_4-$(CH_2)_4$-PPh_2-[Ru(η^6-*p*-cymene)Cl_2(pta)] (right).

http://dx.doi.org/10.5772/67781

centered (LC) or involve both the metal and the ligands: metal-to-ligand charge transfer (MLCT) (for readily oxidized metal ions and ligands with low-lying acceptor orbitals), or ligand-to-metal charge transfer (LMCT) (for readily reduced metal ions with strong donor ligands) (**Figure 5**).

A lot of research is carried out in the delivery of small molecules, which can act as second messengers and transmit signals into cells, for example, NO, carbon monoxide (CO), and hydrogen sulfide (H_2S). Photoactive Pt (IV) diazido complexes also offer potential dual mode activity; excited singlet and triplet states can release reactive or biologically active ligands and form Pt (II) species which can bind to DNA. The introduction of extended conjugation into the amine ligands of square–planar Pt (II) complexes has allowed two-photon activation of ligand exchange using red and near-infrared (NIR) light. The wavelength for two-photon activation of *cis*-[$PtCl_2(MOPEP)_2$], where MOPEP is the π-conjugated ligand 4-[2-(4 methoxyphenyl)ethynyl]pyridine, is shorter than twice the single-photon absorption wavelength [110].

Another important optical phenomenon is "upconversion luminescence," which is discussed here.

3.1. Upconversion luminescence

It is a nonlinear optical phenomenon, which absorb two or more photons and emit one photon. Compared with traditional luminescent materials, upconversion nanostructures have many advantages, such as weak background interference, long lifetime, low excitation energy, and strong tissue penetration, which are used in bioimaging and sensing. Similarly producing shorter wavelength light from longer wavelength irradiation involves the use of upconverting nanoparticles. For example: YF3 doped with lanthanide ions ($Yb^{3+}$and Tm^{3+}). Lanthanide-doped upconversion nanoparticles are used to mediate nitric oxide (NO) release from Roussin's black salt anion [$Fe_4S_3(NO)_7$]$^-$ in NIR light from a simple diode laser operating at 980 nm [110]. Cr (III) sensitizers around a central Er (III) acceptor also favor efficient nonlinear energy transfer and upconversion luminescence [111].

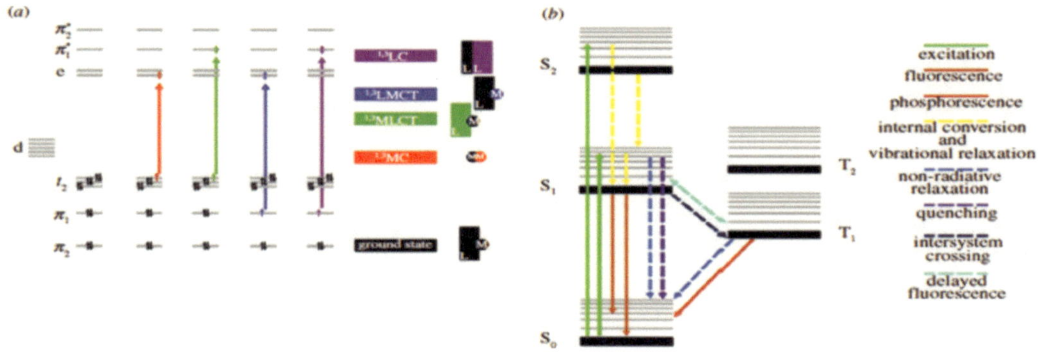

Figure 5. Schematic representation of the orbital and excited state diagram for (d^6) metal complex. Spin is represented by arrows (↑↓) for electronic transitions. (a) Spin up is represented for electronic transition in singlet state whereas spin down is represented for electronic transition in triplet state. (b) Jablonski diagram.

3.2. Imaging and binding of photo-triggered DNA

The simple and powerful strategy for selective destruction of cancer cells is to target the metal complexes to the tumor cells by photoactivation. Peptides releases the aqua species, $[(\eta6\text{-}p\text{-cym})Ru(bpm)(H_2O)]^{2+}$in the visible range which bind to DNA. The other example is the cyclometalated iridium (III) polypyridine indole complexes, which have the intense luminescence (λ_{em} = 540–616 nm, τ = 0.13–5.15 μs) [112]. Another interesting feature of these complexes is that they can deliver CO in the body [113], for example, MnI carbonyl complex $[Mn(pqa)(CO)3]^{+}$ (pqa =(2-pyridylmethyl)(2-quinolylmethyl)amine) [114] and manganese complexes [115]. The release of CO from these complexes is visibly monitored by time-resolved IR spectroscopy [116]. This property is also used to deliver other biologically active species also. $[Rh(bpy)2(chrysi)]^{3+}$ targets single-base mismatches in DNA by noncovalent binding in UV/visible region. As there is deficiency of mismatch repair in cancer cells, this technique can be used to detect the cancer cells [117, 118]. The other luminescent N-heterocylic carbene (NHC) cyclometalated platinum(II) complexes, which are localized in cytoplasmic structures, do not interact with nucleotides [119].

4. Conclusions

Unfortunately, like the macro organometallic complexes, the nanoparticles also carry some serious adverse effects. Though the adverse effects of nanoparticles depend on individual factors such as genetics, existing disease conditions, exposure, nanoparticle chemistry, size, shape, agglomeration state, and electromagnetic properties, the key to understanding the toxicity of nanoparticles is their size. Thus, it is very essential to understand the basic nature, structure, and the photophysics behind these particles. Nanoparticles are smaller than mammalian cells and cellular organelles, which allows them to penetrate these biological structures and disrupt their normal function. Nanoparticles are effective in glycoma treatment. This brain cancer is particularly difficult to treat as neurosurgery is ineffective, while chemotherapy suffers from the inability of therapeutics to cross the blood. Although the lack of self-error-correcting mechanism result in defect sites in these nanostructures, the high efficiency and relative simplicity of the novel approach demonstrates the potential power of using irreversible covalent bonds to generate adverse range of shape-persistent and robust nanostructures that is likely to enrich the repertoire of self-assembled nanomaterials and multidrug delivery. Finally, toxicity of nanoparticles could also be potentially utilized to destroy the cancer cells. Bioorganometallic compounds offer hope in the fight against the deadly diseases such as Malaria, HIV/AIDS, and EVD that have continued to devastate humans. There are expected challenges in this area of collaborative research as organometallic compounds are ideally synthesized under inert atmosphere in the absence of oxygen and water. These challenges are not too difficult to surmount, we therefore implore researchers to orient more into this relatively new multidisciplinary research area in the search for novel and potent anticancer and other drug candidates with reduced side effects, which can be a great service to the mankind.

Acknowledgements

The author acknowledges the financial assistance by the DST WOS-A (CS-1005/2014). The author is also thankful to her mentor Dr. G. Narahari Sastry, Head, Center for Molecular Modeling for the support.

Author details

Ruby Srivastava

Address all correspondence to: amitruby1@gmail.com

CSIR-Indian Institute of Chemical Technology, Hyderabad, Telangana, India

References

[1] Jakupec MA, Galanski M, Arion VB, Hartinger CG, Keppler BK. Antitumour metal compounds: More than theme and variations. Dalton Transactions. 2008;**2**:183-194. DOI:10.1155/2015/859730

[2] Dyson PJ, Sava G. Metal-based antitumour drugs in the post genomic area. Dalton Transactions. 2006;**16**:1929-1933. DOI:10.1039/c1dt10522a

[3] Wang D, Lippard S J. Cellular processing of platinum anticancer drugs. Nature Reviews Drug Discovery. 2005;**4**:307-320. DOI:10.1038/nrd1691

[4] Lippert B. Cisplatin, Chemistry and Biochemistry of a Leading Anticancer Drug; Zurich, Switzerland: Verlag Helvetica Chimica Acta; 1999. p. 563. DOI:10.1038/nrd1691

[5] Gasser G, Ott I and Metzler-Nolte N. Organometallic anticancer compound. Journal of Medicinal Chemistry. 2011;**54**:3-25. DOI: 10.1021/jm100020w

[6] Köpf-Maier P, Köpf H. Organometallic Anticancer Compounds. Drugs Future. 1986;**11**: 297-320. DOI:10.1021/jm100020w

[7] Köpf-Maier P, Köpf H, Neuse EW. Ferrocenium salts; the first antineoplastic iron compounds. Angewandte Chemie, International Edition England. 1984;**23**:456-457. DOI:10.1021/jm100020w

[8] Köpf-Maier P, Köpf H. Non-platinum-group metal antitumor agents: History, current status, and perspectives. Chemical Reviews. 1987; **89**:1137-1152. DOI: 10.1021/cr00081a012

[9] Köpf-Maier P, Köpf H. Transition and main-group metal cyclopentadienyl complexes: preclinical studies on a series of antitumor agents of different structural type. Structural Bonding [Berlin]. 1988;**70**:105-185

[10] Dale LD, Tocher JH, Dyson TM, Edwards DI, Tocher DA. Studies on DNA damage and induction of SOS repair by novel multifunctional bioreducible compounds. II. A metronidazole adduct of a ruthenium-arene compound. Anti-Cancer Drug Design. 1992;**7**:3-14. DOI: 10.2337/dc08-1913

[11] Melchart M, Sadler PJ. Ruthenium arene anticancer complexes. In Jaouen G, editor. Bioorganometallics. Weinheim, Germany: Wiley-VCH; 2006. pp. 39-64

[12] Peacock AFA, Sadler PJ. Medicinal organometallic chemistry: Designing metal arene complexes as anticancer agents. Chemistry, an Asian Journal. 2008;**3**:1890-1899. DOI: 10.1002/asia.200800149

[13] Dougan SJ, Sadler PJ. The design of organometallic ruthenium arene anticancer agents. Chimia. 2007;**61**:704-715. DOI: http://dx.doi.org/10.2533/chimia

[14] Süss-Fink G. Arene ruthenium complexes as anticancer agents. Dalton Transactions. 2010;**39**:1673-1688. DOI: 10.1002/1521-3773

[15] Peacock AFA, Habtemariam A, Fernandez R, Walland V, Fabbiani FPA, Parsons S, Aird RE, Jodrell DI, Sadler PJ. Tuning the reactivity of osmium[II] and ruthenium[II] arene complexes under physiological conditions. Journal of the American Chemical Society. 2006;**128**:1739-1748. DOI: 10.1021/ja055886r

[16] Peacock, AFA, Habtemariam A, Moggach SA, Prescimone A, Parsons S, Sadler P J. Chloro half-sandwich osmium [II] complexes: Influence of chelated N,N-ligands on hydrolysis, guanine binding, and cytotoxicity. Inorganic Chemistry. 2007;**46**:4049-4059. DOI: 10.1021/ic062350d

[17] Peacock AFA, Parsons S, Sadler PJ. Tuning the hydrolytic aqueous chemistry of osmium arene complexes with N, O-chelating ligands to achieve cancer cell cytotoxicity. Journal of the American Chemical Society. 2007;**129**:3348-3357. DOI: 10.1021/ja068335p

[18] Dorcier A, Ang WH, Bolãno S, Gonsalvi L, Juillerat-Jeannerat L, Laurenczy G, Peruzzini M, Phillips AD, Zanobini F, Dyson PJ. In vitro evaluation of rhodium and osmium RAPTA analogues: The case for organometallic anticancer drugs not based on ruthenium. Organometallics. 2006;**25**:4090-4096. DOI: 10.1021/om060394o

[19] Liu Z, and Sadler PJ. Organoiridium complexes: Anticancer agents and catalysts. Accounts of Chemical Research. 2014;**47**(4):1174-1185. DOI: 10.1021/ar400266c

[20] Scharwitz MA, Ott I, Geldmacher Y, Gust R, Sheldrick WS. Cytotoxic half-sandwich rhodium [III] complexes: Polypyridyl ligand influence on their DNA binding properties and cellular uptake. Journal of Organometallics Chemistry. 2008;**693**:2299-2309. DOI: 10.1016/*j*.jorganchem.2008.04.002

[21] Lau JSY, Lee PK, Tsang KHK, Ng CHC, Lam YW, Cheng SH, Lo KKW. Luminescent cyclometalated iridium[III] polypyridine indole complexes;synthesis, photophysics, electrochemistry, protein-binding properties, cytotoxicity, and cellular uptake. Inorganic Chemistry. 2009;**48**:708-718. DOI:10.1007/s11426-010-4120-*y*

[22] Zhang J, Vittal JJ, Henderson W, Wheaton JR, Hall IH, Hor TSA, Yan YK. Tricarbonylrhenium [I] complexes of phosphine-derivatized amines, amino acids and a model peptide: Structures, solution behavior and cytotoxicity. Journal of Organometallics Chemistry. 2002; **650**:123-132. DOI:10.4172/2155-9821.1000141

[23] Wang W, Yan YK, Hor TSA, Vittal JJ, Wheaton JR, Hall IH. Synthesis, X-ray structures, and cytotoxicity of rhenium[I] carbonyl 2-[dimethylamino]ethoxide complexes. Polyhedron 2002;**21**:1991-1999. DOI: 10.1016/s0277-5387[02]01045-8

[24] Yan YK, Cho SE, Shaffer KA, Rowell JE, Barnes BJ, Hall IH. Cytotoxicity of rhenium[I] alkoxo and hydroxo carbonyl complexes in murine and human tumor cells. Pharmazie 2000;**55**:307-313. DOI:10.1007/s11010-014-2201-5

[25] Melaiye A, Sun Z, Hindi, K, Milsted A, Ely D, Reneker DH, Tessier CA, Youngs WJ. Silver[I]-imidazole cyclophane gem-diol complexes encapsulated by electrospun tecophilic nanofibers: Formation of nanosilver particles and antimicrobial activity. Journal of the American Chemical Society. 2005;**127**:2285-2291. DOI:10.1021/ja040226s

[26] Kascatan-Nebioglu A, Melaiye A, Hindi, K, Durmus S, Panzner MJ, Hogue LA, Mallett RJ, Hovis CE, Coughenour M, Crosby SD, Milsted A, Ely DL, Tessier CA, Cannon CL, Youngs WJ. Synthesis from caffeine of a mixed N-heterocyclic carbene-silver acetate complex active against resistant respiratory pathogens. Journal of Medicinal Chemistry. 2006;**49**:6811-6818. DOI:10.1021/jm060711t

[27] Reddy VD, Dayal D, Cosenza SC, Reddy MVR, Pearl WC Jr, Adams RD. Glycal-ruthenium carbonyl clusters: syntheses, characterization, and anticancer activity. Journal of Organometallics Chemistry. 2009;**694**:959-967. DOI:10.1016/j.jorganchem.2008.11.025

[28] Niesel J, Pinto A, N'Dongo HWP, Merz K, Ott I, Gust R, Schatzschneider U. Photoinduced CO release, cellular uptake and cytotoxicity of a tris[pyrazolyl] methane manganese tricarbonyl complex. Chemical Communications. 2008;**15**:1798-1800. DOI: 10.1039/B719075A

[29] Buriez O, Heldt JM, Labbé E, Vessières A, Jaouen G, Amatore C. Reactivity and antiproliferative activity of ferrocenyl-tamoxifen adducts with cyclodextrins against hormone-independent breast-cancer cell lines. Chemistry. 2008;**14**(27):8195-203. DOI: 10.1002/chem.200800507

[30] Patra M, Gasser G, Pinto A, Merz K, Ott I, Bandow JE, Metzler-Nolte N. Synthesis and biological evaluation of chromium bioorganometallics based on the antibiotic Platensimycin lead structure. ChemMedChem. 2009;**4**:1930-1938. DOI: 10.1002/cmdc.200900347

[31] Patra M, Gasser G, Wenzel M, Merz K, Bandow JE, Metzler-Nolte N. Synthesis and biological evaluation of ferrocene-containing bioorganometallics inspired by the antibiotic Platensimycin lead structure. Organometallics. 2010;**29**:4312-4319. DOI: 10.1021/om100614c

[32] Patra M, Gasser G, Wenzel M, Merz K, Bandow JE, Metzler-Nolte N. Sandwich and half-sandwich derivatives of Platensimycin: Synthesis and biological evaluation, Organometallics. 2012;**31**(16):5760-5771. DOI: dx.doi.org/10.1021/om201146c

[33] Bassetti S, Hu J, Agostino Jr. RB, Sherertz RJ. Prolonged antimicrobial activity of catheter containing chlorhexidine silver sulfadiazine extends protection against catheter infection in vivo. Antimicrobial Agents and Chemotherapy. 2001;**45**(5):1535. DOI:10.1128/AAC.45.5.1535-1538.2001

[34] Clarke MJ. Ruthenium metallopharmaceuticals. Coordination Chemistry. 2003;**236**:299-233. DOI: 10.1186/s12951-015-0129-x

[35] Rafique S, Idrees M, Nasim A, Akbar H, and Athar A. Transition metal complexes as potential therapeutic agents. Biotechnology and Molecular Biology Reviews. 2010;**5**(2):38-45. DOI: http://www.academicjournals.org /BMBR

[36] Singh RV, Chaudhary A. Biologically relevant tetra azamacrocyclic complexes of manganese. Spectral, antimicrobial, antifertility, and anti-inflammatory approach. Journal of Inorganic Biochemistry. 2004;**98**(11):1712-1721. DOI: 10.1016/j.jinorgbio.2004.07.007

[37] Kemp MM, Kumar A, Mousa S, Dyskin E, Yalcin M, Ajayan P, Linhardt RJ, Mousa SA. Gold and silver nanoparticles conjugated with heparin derivative possess anti-angiogenesis properties. Nanotechnology. 2009;**20**(45):455104. DOI:10.1088/0957-4484/20/45/455104

[38] Nash P, Clegg DO. Proriatic arthritis therapy: NSAID and traditional DMARDS. Annals of the Rheumatic Diseases. 2005;**64**:74-77. DOI: 10.1136/ard.2004.030783

[39] Pedersen MO, Larsen A, Pedersen DS, Stoltenberg M, Penkova M. Metallic gold treatment reduces proliferation of inflammatory cells, increases expression of VEGF and FGF, and stimulates cell proliferation in the subventricular zone following experimental traumatic brain injury. Histology Histopathology. 2009;**5**:573-586. DOI:10.14670/HH-24.573

[40] Mosi R, Seguin B, Cameron B, Amankawa L. Mechanistic studies on AMD6221: A ruthenium based nitric oxide scavenger. Biochemical and Biophysical Research Communications. 2009;**292**:519-529. DOI: 10.1006/bbrc.2002.6685

[41] Spasojevic I, Batinic-Harberle I, Reboculus. Electrostatic contribution in the catalysis of O_2- dismutation by superoxide dismutase mimics. Journal of Medicinal Chemistry. 2003;**278**:6831-6837. DOI:10.1074/jbc.M211346200

[42] Bagonza J, Rutebemberwa E, Bazeyo W. Adherence to anti diabetic medication among patients with diabetes in eastern Uganda; a cross sectional study. BMC Health Services Research. 2015;**15**:168-175. DOI: 10.1186/s12913-015-0820-5

[43] Balk EM, Tatsioni A, Lichtenstein AH, Lau J, Pittas AG. Effect of chromium supplementation on glucose metabolism and lipids. A systematic review of randomized controlled trials. Diabetes Care. 2007;**30**(8):2154-2163. DOI: https://doi.org/10.2337/dc06-0996

[44] Mehdi MZ, Pandey SK, Theberge JF, Srivastava AK. Insulin signal mimicry as a mechanism for the insulin-like effects of vanadium. Cell Biochemical and Biophysics. 2006;**44**(1):73-81. DOI: 10.1385/CBB:44:1:073

[45] Sun Q, van Dam RM, Willett WC, Hu FB. Prospective study of zinc intake and risk of type 2 diabetes in women. Diabetes Care. 2009;**32**(4):629-634. DOI: 10.2337/dc08-1913

[46] Renfrew AK, Phillips AD, Tapavicza E, Scopelliti R, Rothlisberger U, Dyson PJ. Tuning the efficacy of ruthenium-[II]-Arene [RAPTA] antitumor compounds with fluorinated arene ligands. Organometallics 2009;**28**:5061-5071. DOI:10.1021/om900345n

[47] Leonidova A. Gasser G. The underestimated potential of organometallic rhenium complexes as anticancer agents. Biology. 2014;**9**(10):2180-2193. DOI: 10.1021/cb500528c

[48] Ali Nazif M, Bangert JA, Ott I, Gust R, Stoll R, Sheldrick WS. Dinuclear organoiridium [III] mono- and bis-intercalators with rigid bridging ligands: synthesis, cytotoxicity and DNA binding. Journal of Inorganic Biochemistry. 2009;**103**(10):1405-1414. DOI: 10.1016/j.jinorgbio.2009.08.003

[49] Kostrhunova H, Florian J, Novakova O, Peacock AFA, Sadler PJ, Brabec V. DNA interactions of monofunctional organometallic osmium[II] antitumor complexes in cell-freemedia. Journal of Medicinal Chemistry. 2008;**51**:3635-3643. DOI:10.1021/jm701538w

[50] Van Rijt, SH, Peacock AFA, Johnstone RDL, Parsons S, Sadler PJ. Organometallic osmium[II] arene anticancer complexes containing picolinate derivatives. Inorganic Chemistry. 2009; **48**:1753-1762. DOI:10.1021/ic8020222

[51] Schmid WF, John RO, Arion VB, Jakupec MA, Keppler BK. Highly antiproliferative ruthenium[II] and osmium[II] arene complexes with paullone-derived ligands. Organometallics. 2007;**26**:6643-6652. DOI:10.1039/c5dt02410b

[52] Dorcier A, Dyson PJ, Gossens C, Rothlisberger U, Scopelliti R, Tavernelli I. Binding of organometallic ruthenium[II] and osmium[II] complexes to an oligonucleotide: A combined mass spectrometric and theoretical study. Organometallics. 2005;**24**:2114-2123. DOI:10.1002/ejic.201100250

[53] Osella D, Mahboobi H, Colangelo D, Cavigiolio G, Vessieres A, Jaouen G. FACS analysis of oxidative stress induced on tumour cells by SERMs. Inorganica Chimica Acta. 2005;**358**:1993-1998. DOI: http://dx.doi.org/10.1016/j.ica.2004.11.027

[54] Arezki A, Brule E, Jaouen G. Synthesis of the first ferrocenyl derivatives of curcuminoids. Organometallics. 2009;**28**:1606-1609. DOI:10.1039/c0md00231c

[55] Top S, Thibaudeau C, Vessieres A, Brule E, Le Bideau F, Joerger JM, Plamont MA. Samreth S, Edgar A, Marrot J, Herson P, Jaouen G. Synthesis and structure activity relationship of organometallic steroidal androgen derivatives. Organometallics. 2009;**28**:1414-1424. DOI:10.1021/om800698y

[56] Payen O, Top S, Vessieres A, Brule E, Plamont MA, McGlinchey M J, Mueller-Bunz H, Jaouen G. Synthesis and structure-activity relationships of the first ferrocenyl-aryl-hydantoin derivatives of the nonsteroidal antiandrogen nilutamide. Journal of Medicinal Chemistry. 2008;**51**:1791-1799. DOI:10.1021/jm701264d

[57] Spencer J, Mendham AP, Kotha AK, Richardson SCW, Hillard EA, Jaouen G, Male L, Hursthouse MB. Structural and biological investigation of ferrocene-substituted 3-methylidene-1,3-dihydro-2H-indol-2-ones. Dalton Transactions. 2009;**6**:918-921. DOI: 10.1039/B816249B

[58] Plazuk D, Vessieres A, Hillard EA, Buriez O, Labbe E, Pigeon P, Plamont MA, Amatore C, Zakrzewski J, Jaouen GA. [3]Ferrocenophane polyphenol showing a remarkable antiproliferative activity on breast and prostate cancer cell lines. Journal of Medicinal Chemistry. 2009;**52**:4964-4967. DOI:10.1021/jm900297x

[59] Raubenheimer HG, Cronje S. Carbene complexes of gold: Preparation, medical application and bonding. Chemical Society Reviews. 2008;**37**:1998-2011. DOI:10.1039/b708636a

[60] Jones GB, Mathews JE. Tricarbonyl arene chromium[0] based antitumor agents. Bioorganic Medical Chemistry Letters. 1995;**5**:93-96. DOI:10.1016/j.ejphar.2010.01.011

[61] Neundorf I, Hoyer J, Splith K, Rennert R, N'Dongo HWP, Schatzschneider U. Cymantrene conjugation modulates the intracellular distribution and induces high cytotoxicity of a cell-penetrating peptide. Chemical Communications. 2008;**21**:5604-5606. DOI: 10.1021/jm100020w

[62] Murthy N, Rao AR, Sastry GN. Aromatase inhibitors: A new paradigm in breast cancer treatment. Current Medicinal Chemistry. Anti-Cancer Agents. 2004;**4**:523-534. DOI: 10.1007/s00044-011-9688-z

[63] Srinivas E, Murthy JN, Ram Rao AR, Sastry GN. Recent advances in molecular modeling and medicinal chemistry aspects of phospho-glycoprotein. Current Drug Metabolism. 2006;**7**:205-217. DOI:10.2174/138920006775541534

[64] Kulkarrni RG, Achaiah G, Sastry GN. Novel targets for anti-inflammatory and antiarthritic agents. Current Pharmaceutical Design. 2006;**12**(19):2437-2454. PMID:16842190

[65] Kamal A, Naseer M, Khan A, Reddy KS, Rohini K, Sastry GN, Sateesh B, Sridhar B. Synthesis, structure analysis, and antibacterial activity of some novel 10-substituted 2-(4-piperidyl/phenyl)-5,5-dioxo[1, 2, 4] triazolo[1, 5-b][1, 2, 4] benzothiadiazine derivatives. Bioorganic & Medicinal Chemistry Letters. 2007;**17**:5400-5405. DOI: 10.1016/j.bmcl.2007.07.043

[66] Kamal A, Rajender, Reddy DR, Reddy MK, Balakishan G, Shaik TB, Chourasia M, Sastry GN. Remarkable enhancement in the DNA-binding ability of C2-fluoro substituted pyrrolo[2, 1-c][1, 4]benzodiazepines and their anticancer potential. Bioorganic & Medicinal Chemistry. 2009;**17**:1557-1572. DOI:10.1016/j.bmc.2008.12.068

[67] Kamal A, Bharathi EV, Ramaiah MJ, Dastagiri D, Surendranadha Reddy J, Viswanath, A, Sultana F, Pushpavalli SNCVL, Bhadra MP, Srivastava HK, Sastry GN, Juvekar A, Sen S, Zingde S. Quinazolinone linked pyrrolo[2, 1-c] [1, 4] benzodiazepine (PBD) conjugates: Design, synthesis and biological evaluation as potential anticancer agents. Bioorganic & Medicinal Chemistry. 2010;**18**:526-542. DOI:10.1002/jcc.21277

[68] Kamal A, Shetti RVCRNC, Ramaiah MJ, Swapna P, Reddy KS, Mallareddy A, Narasimha Rao MP, Chourasia M, Sastry GN, Juvekar A, Zingde S, Sarma P, Pushpavallib SNCVL, Bhadra MP. Carbazole–pyrrolo[2, 1-c][1, 4]benzodiazepine conjugates: design, synthe-

sis, and biological evaluation. Medicinal Chemistry Communications. 2011;**2**:780-788. DOI: 10.1039/C1MD00072A

[69] Bohari MH, Srivastava HK, Sastry GN. Analogue-based approaches in anti-cancer compound modelling: The relevance of QSAR models, Bohari et al. Organic and Medicinal Chemistry Letters. 2011;**1**:1-12. DOI: 10.1186/2191-2858-1-3

[70] Sreshty AS, Surolia A, Sastry GN, Suryanarayana Murty U. Deorphanization of Malonyl CoA:ACP transacylase drug target in plasmodium falciparum (PfFabD) using bacterial antagonists: A 'Piggyback' approach for antimalarial drug discovery. Molecular Informatics. 2012;**31**: 281-299. DOI: 10.1002/minf.201100051

[71] Venkatesh R, Ramaiah MJ, Gaikwad HK, Janardhan S, Bantu RS, Nagarapu L, Sastry GN, Ganesh AR, Bhadra MP. Luotonin-A based quinazolinones cause apoptosis and senescence via HDAC inhibition and activation of tumor suppressor proteins in HeLa Cells. European Journal of Medicinal Chemistry. 2015;**94**:87-101. DOI: http://dx.doi.org/10.1016/j.ejmech.2015.02.057

[72] Ponti E, Sabbioni B. Munaro et al. Genotoxicity and morphological transformation induced by cobalt nanoparticles and cobalt chloride: An in vitro study in Balb/3T3 mouse fibroblasts. Mutagenesis. 2009;**24**:439-445. DOI: 10.1093/mutage/gep027

[73] Dey S, Bakthavatchalu V, Tseng MT et al. Interactions between SIRT1 and AP-1 reveal a mechanistic insight into the growth promoting properties of alumina [Al_2O_3] nanoparticles in mouse skin epithelial cells. Carcinogenesis. 2008;**29**(10):1920-1929. DOI: 10.1093/carcin/bgn175

[74] Sangiliyandi G, Han JW, Eppakayala V, Jeyaraj M, Kim JH. Cytotoxicity of biologically synthesized silver nanoparticles in MDA-MB-231 Human Breast Cancer Cells BioMed Research International Volume 2013;1-10. DOI:10.1155/2013/535796

[75] Simon M, Barberet P, Delville MH, Moretto P, and Seznec H. Titanium dioxide nanoparticles induced intracellular calcium homeostasis modification in primary human keratinocytes. Towards an in vitro explanation of titanium dioxide nanoparticles toxicity. Nanotoxicology. 2011;**5**(2):125-139. DOI:10.3109/17435390.2010.502979

[76] Blasiak B, Frank CJM van Veggel, Tomanek B. Applications of nanoparticles for MRI cancer diagnosis and therapy. Journal of Nanomaterials. 2013;**2013**:1-12. DOI: http://dx.doi.org/10.1155/2013/148578

[77] Meyer MHF, Stehr M, Bhuju S et al. Magnetic biosensor for the detection of Yersinia pestis. Journal of Microbiological Methods. 2007;**68**(2):218-224. DOI: http://dx.doi.org/10.1155/2013/148578

[78] Kirsch JE. Basic principles of magnetic resonance contrast agents. Topics in Magnetic Resonance Imaging. 1991;**3**(2):1-18. DOI:10.1155/2013/148578

[79] Cai W, Gao T, Hong H, Sun J. Applications of gold nano-particles in cancer. Nanotechnology, Science and Applications. 2008;**1**:17-32. DOI: https://doi.org/10.2147/NSA.S3788

[80] Duguet E, Vasseur S, Mornet S, Devoisselle JM. Magnetic nanoparticles and their applications in medicine. Nanomedicine. 2006;**1**(2):157-168. DOI:10.2217/17435889.1.2.157

[81] Elblbesy M. Hemocompatibility of albumin nanoparticles as a drug delivery system—An *in vitro* study. Journal of Biomaterials and Nanobiotechnology. 2016;**7:**64-71. DOI:10.4236/jbnb.2016.72008

[82] Fahmy TM, Fong PM, Goyal A, Saltzman WM. Targeted for drug delivery. Materials Today. 2005;**8(8)**:18-26. DOI: http://dx.doi.org/10.1016/S1369-7021[05]71033-6

[83] Fahmy TM, Samstein RM, Harness CC, Saltz-man WM. Surface modification of biodegradable polyesters with fatty acid conjugates for improved drug targeting. Biomaterials. 2005;**26**(28):5727-5736. DOI:10.1016/j.biomaterials.2005.02.025

[84] Couvreur P, Barratt G, Fattal E, Legrand P, Vauthier C. Nanocapsule technology: A review. Critical Reviews in Therapeutic Drug Carrier Systems. 2002;**19**(2):99-134. DOI: 10.5681/apb.2012.007

[85] Roger J, Pons JN, Massart R, Halbreich A, Bacri JC. Some biomedical applications of ferrofluids. The European Physical Journal Applied Physics. 1999;**5**(3):321-325. DOI: http://dx.doi.org/10.1155/2013/148578

[86] Alexis F, Rhee JW, Richie JP, Radovic-Moreno AF, Langer R, Farokhzad OC. New frontiers in nanotechnology for cancer treatment. Urologic Oncology. 2008;**26**(1):74-85. DOI:http://doi.org/10.1016/j.urolonc.2007.03.017

[87] Zhang L, Gu FX, Chan JM, Wang AZ, Langer RS, Farokhzad OC. Nanoparticles in medicine: Therapeutic applications and developments. Clinical Pharmacology and Therapeutics. 2008;**83**(5):761-769. DOI:10.1038/sj.clpt.6100400

[88] Phanapavudhikul P, Shen S, Ng WK, Tan RBH. Formulation of Fe_3O_4/acrylate co-polymer nanocomposites as potential drug carriers. Drug Delivery. 2008;**15**(3):177-183. DOI: http://dx.doi.org/10.1080/10717540801952597

[89] Arap W, Pasqualini R, Ruoslahti E. Cancer treatment by targeted drug delivery to tumor vasculature in a mouse model. Science. 1998;**279**(5349):377-380. DOI: 10.1126/science.279.5349.377

[90] Dagar S, Sekosan M, Lee BS, Rubinstein I, Ony H. VIP receptors as molecular targets of breast cancer: Implications for targeted imaging and drug delivery. Journal of Controlled Release. 2001;**74**(1-3):129-134. DOI:10.1016/S0168-3659[01]00326-1

[91] Sousa F, Mandal S, Garrovo C et al. Functionalized gold nanoparticles: A detailed in vivo multimodal microscopic brain distribution study. Nanoscale. 2010;**2**(12):2826-2834. DOI: http://dx.*doi*.org/10.1039/c0nr00345j

[92] Wen PY, Kesari S. Malignant gliomas in adults. New England Journal of Medicine. 2008;**359**(5):492-507. DOI:10.1056/NEJMra0708126

[93] Ullrich NJ, Pomeroy SL. Pediatric brain tumors. Neurologic Clinics. 2003;**21**(4):897-913. DOI:10.1016/S0733-8619(03)00014-8

[94] Hernández-Pedro NY, Rangel-López E, Magãna-Maldonado R et al. Application of nanoparticles on diagnosis and therapy in gliomas. BioMed Research International. 2013;**2013**:1-20. DOI: http://dx.doi.org/10.1155/2013/351031

[95] Park JY, Baek MJ, Choi ES et al. Paramagnetic ultrasmall gadolinium oxide nanoparticles as advanced T1 MRI contrast agent: Account for large longitudinal relaxivity, DOI:10.1021/nn900761s

[96] Lefebure S, Dubois E, Cabuil V, Neveu S, Massart R. Monodisperse magnetic nanoparticles: Preparation and dispersion in water and oils. Journal of Materials Research. 1998;**13**(10):2975-2981. DOI:10.1557/JMR.1998.0407

[97] Mejías R, Pèrez-Yagüe S, Roca AG et al. Liver and brain imaging through dimercaptosuccinic acid-coated iron oxide nanoparticles. Nanomedicine. 2010;**5**(3):397-408. DOI:10.2217/nnm.10.15

[98] Siegal T, Horowitz A, Gabizon A. Doxorubicin encapsulated in sterically stabilized liposomes for the treatment of a brain tumor model: Biodistribution and therapeutic efficacy. Journal of Neurosurgery. 1995;**83**(6):1029-1037. DOI:10.3171/jns.1995.83

[99] Kircher MF, de la Zerda A, Jokerst JV et al. A brain tumor molecular imaging strategy using a new triple-modality MRI-photoacoustic-Raman nanoparticle. Nature Medicine. 2012;**18**(5):829-834. DOI:10.1038/nm.2721

[100] Khlebtsov N, Dykman L. Biodistribution and toxicity of engineered gold nanoparticles: A review of in vitro and in vivo studies. Chemical Society Reviews. 2011;**40**(3):1647-1671. DOI:10.1038/nm.2721

[101] De Jong WH, Hagens WI, Krystek P, Burger MC, Sips AJAM, Geertsma RE. Particle size-dependent organ distribution of gold nanoparticles after intravenous administration. Biomaterials. 2008;**29(12)**:1912-1919. DOI: 10.1016/j.biomaterials.2007.12.037

[102] Sonavane G, Tomoda K, and Makino K. Biodistribution of colloidal gold nanoparticles after intravenous administration: Effect of particle size. Colloids and Surfaces B. 2008;**66**(2):274-280. DOI: 10.1016/j.colsurfb.2008.07.004

[103] Xia H, Li F, Hu X, Park W, Wang S, Jang Y, Du Y, Baik S, Cho S, Kang T, Kim DH, Ling D, Hui KM, Hyeon T. pH-Sensitive Pt nanocluster assembly overcomes cisplatin resistance and heterogeneous stemness of hepatocellular carcinoma. ACS Central Science. 2016;**2**:802–811. DOI: 10.1021/acscentsci.6b00197

[104] Bardhan R, Lal S, Joshi A, Halas N. Theranostic nanoshells: From probe design to imaging and treatment of cancer. Journal of Accounts of Chemical Research. 2011;**44**:936–946. DOI: 10.1021/ar200023x

[105] Scrase TG, O Neill MJ, Peel AJ, Senior PW, Matthews PW, Shi H, Boss SR, Barker PD. Selective lability of Ruthenium(II) arene amino acid complexes. Inorganic Chemistry. 2015;**54**:3118–3124. DOI: 10.1021/ic502051y

[106] Clavel CM, Paunescu E, Nowak-Sliwinska P, Griffioen AW, Scopelliti R, Dyson PJ. Discovery of a highly tumor-selective organometallic ruthenium [II]–arene complex. Journal of Medicinal Chemistry. 2014;**57**:3546–3558. DOI:10.1021/jm5002748

[107] Allardyce CS, Dyson PJ, Ellis DJ, Heath SL. Ligand substitutions between ruthenium–cymene compounds can control protein versus DNA targeting and anticancer activity. Chemical Communications. 2001;1396–1397. DOI:10.1039/b705449a

[108] Ang WH, Daldini E, Juillerat-Jeanneret L, Dyson PJ. Strategy to tether organometallic ruthenium-arene anticancer compounds to recombinant human serum albumin. Inorganic Chemistry. 2007;**46**:9048–9050. DOI:10.1021/ic701474m

[109] Sun RW, Li CK, Ma DL, Yan JJ, Lok CN, Leung CH, Zhu N, Che CM. Stable anticancer gold[III]-porphyrin complexes: Effects of porphyrin structure. Chemistry. 2010;**16**[10]:3097-3113. DOI: 10.1002/chem.200902741

[110] Garcia JV et al. NIR-triggered release of caged nitric oxide using upconverting nanostructured materials. Small. 2011;**8**:3800-3805. DOI:10.1002/smll.201201213

[111] Aboshyan-Sorgho L, Besnard C, Pattison P, Kittilstved KR, Aebischer A, Bünzli J-CG, Hauser A, Piguet C. Near-infrared → visible light upconversion in a molecular trinuclear d–f–d complex. Angewandte Chemie International Edition. 2011;**50**:4108-4112. DOI:10.1002/anie.201100095

[112] Wecksler SR, Mikhailovsky A, Korystov D, Buller F, Kannan R, Tan L-S, Ford PC. Single- and two-photon properties of a dye-derivatized Roussin's red salt ester $[Fe_2[\mu\text{-}RS]_2[NO]_4]$ with a large TPA cross section. Inorganic Chemistry. 2007;**46**:395-402. DOI:10.1021/ic0607336

[113] Schatzschneider U. PhotoCORMs: Light-triggered release of carbon monoxide from the coordination sphere of transition metal complexes for biological applications. Inorganic Chemistry Acta. 2011;**374**:19-23. DOI:10.1016/j.ica.2011.02.068

[114] Gonzalez MA, Yim MA, Cheng S, Moyes A, Hobbs AJ, Mascharak PK. Manganese carbonyls bearing tripodal polypyridine ligands as photoactive carbon monoxide-releasing molecules. Inorganic Chemistry. 2011;**51**:601-608. DOI:10.1021/ic2021287

[115] Pfeiffer H, Rojas A, Niesel J, Schatzschneider U. Sonogashira and 'Click' reactions for the N-terminal and side-chain functionalization of peptides with $[Mn[CO]_3[tpm]]^+$-based CO releasing molecules [tpm = tris[pyrazolyl]methane]. Dalton Transactions. 2009;**22**:4292-4298. DOI:10.1039/b819091g

[116] Huber W, Linder R, Niesel J, Schatzschneider U, Spingler B, Kunz PC. A comparative study of tricarbonylmanganese photoactivatable CO releasing molecules [PhotoCORMs] by using the myoglobin assay and time-resolved IR spectroscopy. European Journal of Inorganic Chemistry. 2012;**19**:3140-3146. DOI:10.1002/ejic.201200115

[117] Zeglis BM, Pierre VC, Barton JK. 2007 Metallo-intercalators and metallo-insertors. Chemical Communications. 2007;**44**:4565-4579. DOI:10.1039/b710949k

[118] Zeglis BM, Barton JK. DNA base mismatch detection with bulky rhodium intercalators: Synthesis and applications. Nature Protocols. 2007;**2**:357-371. DOI:10.1038/nprot.2007.22

[119] Wai-Yin Sun R, Lok-Fung Chow A, Li XH, Yan JJ, Sin-Yin Chui S, Che C-M. Luminescent cyclometalated platinum[II] complexes containing N-heterocyclic carbene ligands with potent in vitro and in vivo anti-cancer properties accumulate in cytoplasmic structures of cancer cells. Chemical Science. 2011;**2**:728-736. DOI:10.1039/c0sc00593b

Chapter 2

Radical Mechanisms in the Metallocenes

Betul Caliskan

Additional information is available at the end of the chapter

http://dx.doi.org/10.5772/intechopen.68952

Abstract

A special class of sandwich complexes is the metallocenes. The best-known members are the metallocenes of the formula $M(C_5H_5)_2$, where M = zirconium, zinc, titanium, hafnium, vanadium, chromium, molybdenum, tungsten, manganese, iron, ruthenium, osmium, cobalt, rhodium, and nickel. Besides the two cyclopentadienyl rings, the metal can have additional ligands depending on its valence state. These species are also called bis(cyclopentadienyl)metal complexes. Bis(cyclopentadienyl) complexes are called metallocenes or sandwich compounds. Metal-centred radical and cyclopentadienyl radical structures can occur in the metallocene. The anion and cation radicals can also be formed by charge transfer transition.

Keywords: metallocenes, sandwich compound, organometallic compounds, cyclopentadienyl complex, π-complexes, σ-complexes, ionic complexes

1. Introduction

The first metallocene to be classified was ferrocene and was discovered simultaneously in 1951 by Kealy and Pauson; Keally and Pauson were trying to synthesize fulvalene by oxidation of a cyclopentadienyl salt with anhydrous $FeCl_3$. However, instead of this, $C_{10}H_{10}Fe$ is obtained [1]. Miller et al. also reported the same iron product obtained from the reaction of cyclopentadiene with iron in the presence of aluminum, potassium or molybdenum oxides [2]. Wilkinson et al. [3] and Fischer and Pfab [4] studied ferrocene (**Figure 1**) and received the Nobel Prize in 1973 for their work on sandwich compounds. In the bond between the iron atom and the carbon atoms in the cyclopentadienyl (Cp) ligands, each of the carbon atoms contributes equally to the bond. They found that each of the carbon atoms contributed equally to the bond between the carbon atoms in the cyclopentadienyl ligands and the iron atom. In this connection, the effective electrons are the electrons in the d-orbitals of the iron

INTECH
open science | open minds

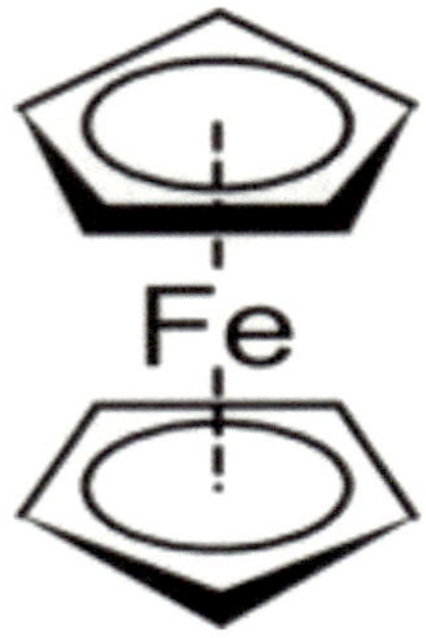

Figure 1. Ferrocene.

and the π-electrons in the p-orbitals of the Cp ligands. This complex is called ferrocene. This kind of construction was called the metolocene. Co and Ni were used as metals in the first metallocenes. Metallocenes of many elements derived mostly from derivatives used instead of cyclopentadienide have been prepared [5].

2. Classification of metallocenes

A metallocene is a compound typically consisting of two cyclopentadienyl anions (Cp, which is C_5H_5) bound to a metal center (M) in the oxidation state II, with the resulting general formula $(C_5H_5)_2M$. For metalocene derivatives, titanocene dichloride and vanadocene dichloride are the best examples. Some metallocenes and their derivatives exhibit catalytic properties, but metallocenes are seldom used industrially. Cationic group 4 metallocene derivatives are concerned with $[Cp_2ZrCH_3]^+$ catalyze olefin polymerization. Metallocenes are a subset of organometallic compounds called sandwich compounds.

In **Figure 1**, the two parallel structures are cyclopentadienyl anions, which are aromaticly stable and show an ordered arrangement.

Ferrocene, $(C_5H_5)_2Fe$ or Cp_2Fe, is systematically called bis(η^5-cyclopentadienyl)iron(II). Due to the ferrocene name, the first found compound, this type of ferrocene-like structure is called metallocene. IUPAC describes a metallocene as a transition metal between two cyclopentadienyl anions which consist of parallel planes with equal bond lengths and strengths in the sandwich structure. The term "hapticity" means that the equivalent of the five carbon atoms of a cyclopentadienyl ring is completely equivalent. It is pronounced as "pentahapto." It is expressed as η^5. There are exceptions, such as uranocene, in which a uranium atom is sandwiched between two cyclooctatetraene rings.

IUPAC defines the term metalocene as a combination of a d-block metal and a sandwich structure. However, the metallocene term expressed by the sign of the *-ocene* is also applied to compounds such as manganocene or titanocene dichloride (Cp_2TiCl_2) in which the aromatic rings are not parallel or to the i non-transition metal compounds (Cp_2Ba) such as barocene in the chemical literature.

Some metallocene complexes of actinides have been reported to have three cyclopentadienyl ligands for a monometallic complex, and all have been reported to bind η^5 [6].

(η^5-C_5H_5)-metallocene complexes can be classified as shown in **Table 1** [7]. Various (η^5-C_5H_5)-metal complexes are also shown in **Figures 2–4**.

Metallocene complexes can also be classified by type [7]:

(A). Parallel (sandwich)

(B). Half-sandwich

(C). Bent or tilted

(D). Multi-decker [8]

(E). More than two Cp ligands

2.1. Sandwich compound

In organometallic chemistry, a sandwich compound is a chemical compound containing a metal attached to two arene ligands via haptic-covalent bonds. The arenes have the formula C_nH_n, substituted derivatives (for example $C_n(CH_3)_n$) and heterocyclic derivatives (for example BC_nH_{n+1}). Since the metal is usually placed between the two rings, it is said to be a "sandwiched." The metallocene is a special sandwich complex class.

Formula	Description
$[(\eta^5\text{-}C_5H_5)_2M]$	Symmetrical, classical "sandwich" structure
$[(\eta^5\text{-}C_5H_5)_2ML_x]$	Bent or tilted Cp rings with additional ligands, L
$[(\eta^5\text{-}C_5H_5)ML_x]$	Only one Cp ligand with additional ligands, L ("piano-stool" structure)

Table 1. Classification of (η^5-C_5H_5)-metal complexes.

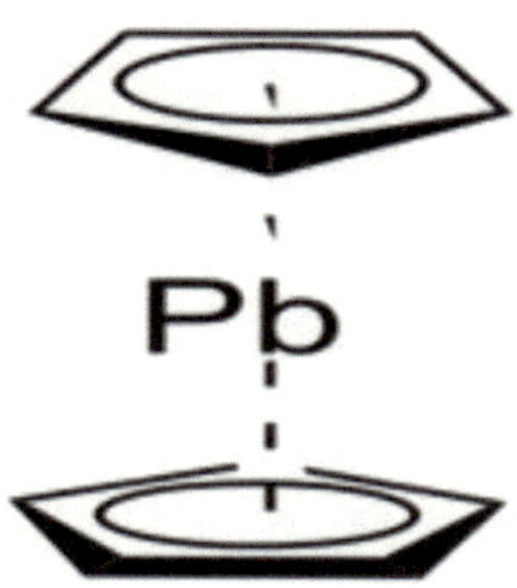

Figure 2. Sandwich or metallocene complexes.

The sandwich compound term entered to the organometallic nomenclature with their studies identifying the structure of ferrocene by X-ray crystallography of Dunitz et al. [9]. The correct structure was proposed several years ago by Robert Burns Woodward and Ernst Otto Fischer separately. The structure helped to explain the unknowns of ferrocene's conformers with an iron atom trapped between two parallel cyclopentadienyl rings. This result showed better the power of X-ray crystallography and accelerated the growth of organometallic chemistry [10]. Ferrocene, a sandwich compound, is shown in **Figure 5**.

Figure 3. Bent or tilted metallocenes.

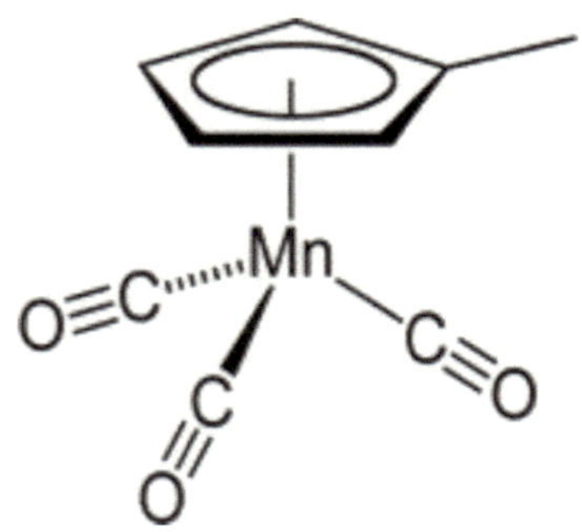

Figure 4. Half-sandwich complexes ("piano-stool" complexes).

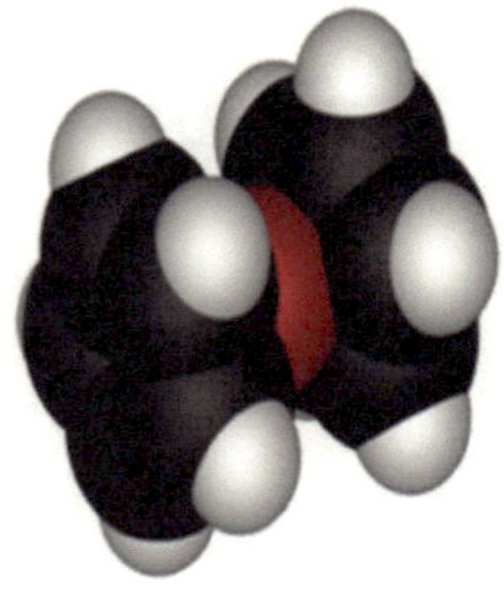

Figure 5. Ferrocene, the archetypal sandwich compound.

The metallocene is generally expressed by the formula $M(C_5H_5)_2$ where M = Cr, Fe, Co, Ni, Pb, Zr, Ru, Rh, Sm, Ti, V, Mo, W, and Zn. Metallosenes are also called bis (cyclopentadienyl) metal complexes.

Apart from the metallocene, the parallel sandwich complexes can be listed as follows. These are not metallocene. However, these are parallel and sandwich compounds.

- **Mixed cyclopentadienyl complexes**: Its general structure is $M(C_5H_5)(C_nH_n)$, such as $Ti(C_5H_5)(C_7H_7)$ and $(C_{60})Fe(C_5H_5Ph_5)$.
- **Bis(benzene)complexes**: Its general structure is $M(C_6H_6)_2$, such as bisbenzene chromium.
- **Bis(cyclooctatetraenyl) complexes**: Its general structure is $M(C_8H_8)_2$, such as $U(C_8H_8)_2$ and $Th(C_8H_8)_2$.
- **Bis(cyclobutadienyl) complexes**: Its general structure is $M(C_4H_4)_2$, such as $Fe(C_4H_4)_2$.

The sandwich compounds can also be formed with inorganic ligands such as $Fe(C_5Me_5)(P_5)$ and $[(P_5)_2Ti]^{2-}$ [11].

2.2. Half-sandwich compounds

It is also called piano-stool compounds. The name of the piano-stool compounds originates from the fact that the structure is likened to a "stool" having a seat with a facial planar organic compounds such as benzene or cyclopentadiene and legs with ligands such as CO or allyl. The metallocenes may have a larger half-sandwich compound family even with a planar organic ligand attached to the surface instead of the two ligands.

2.2.1. *Monometallic half-sandwich compounds*

These compounds are organometallic. The general structure of the half-sandwich compounds consists of the cyclic polyhapto ligand bound to ML_n, where L is a conjugated ligand [12]. $(C_5H_5)TiCl_3$ and $(C_5H_5)Co(CO)_2$ complexes are among the best examples of this group. Half-sandwich compounds according to the cyclic polyhapto ligand can be examined in two categories: $(\eta^5\text{-}C_5H_5)$ piano stool compounds and $(\eta^6\text{-}C_6H_6)$ piano stool compounds.

(A). $(\eta^5\text{-}C_5H_5)$ piano stool compounds: It contains the cyclopentadienyl ligand, $(\eta^5\text{-}C_5H_5)$, as the cyclic polyhapto ligand of the half-sandwich complex. $(\eta^5\text{-}C_5H_5)V(CO)_4$, $(\eta^5\text{-}C_5H_5)Cr(CO)_3H$, $(\eta^5\text{-}C_5H_5)Mn(CO)_3$, $(\eta^5\text{-}C_5H_5)Cr(CO)_3H$, $[(\eta^5\text{-}CH_3C_5H_4)Fe(CO)_3]_+$, $(\eta^5\text{-}C_5H_5)V(CO)_4I$ and $(\eta^5\text{-}C_5H_5)Co(CO)_2$ complexes are among the best examples of this group. $(\eta^5\text{-}C_5H_5)Co(CO)_2$ complex is shown in **Figure 6**.

(B). $(\eta^6\text{-}C_6H_6)$ piano stool compounds: It contains the benzene ring, $(\eta^6\text{-}C_6H_6)$, as the cyclic polyhapto ligand of the half-sandwich complex. The general structure of these compounds consists $(\eta^6\text{-}C_6H_6)ML_3$, where L is typically CO-conjugated ligand and M are metals such as Cr, Mo, W, Mn(I), and Re(I). $(\eta^6\text{-}C_6H_6)$ piano stool complexes are stable compounds with a variety of chemical and material applications and an example of $(\eta^6\text{-}C_6H_6)$ piano stool complex is given in **Figure 7**. The first studies on the $(\eta^6\text{-}C_6H_6)Cr(CO)_3$ compound were carried out in 1957 and 1958 [13–15]. The crystal structure was determined in 1959 [16]. The subsequent studies came in 1987 and 1989 [17, 18]. The ball-and-stick model of methylcyclopentadienyl manganese tricarbonyl, a "piano stool" compound and $CpFe(CO)_2I$, an example of an unsymmetrical piano stool complex, are shown in **Figures 8** and **9**, respectively.

2.2.2. Dimetallic half-sandwich compounds

Compounds such as the cyclopentadienyliron dicarbonyl dimer and cyclopentadienylmolybdenumtricarbonyl dimer can be considered a special case of half-sandwiches, except that they are dimetallic [17]. A structurally related species is $[Ru(C_6H_6)Cl_2]_2$.

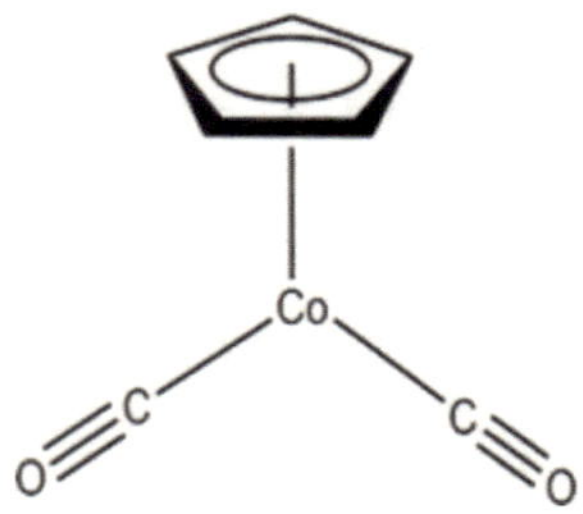

Figure 6. (η^5-C_5H_5) piano stool compound.

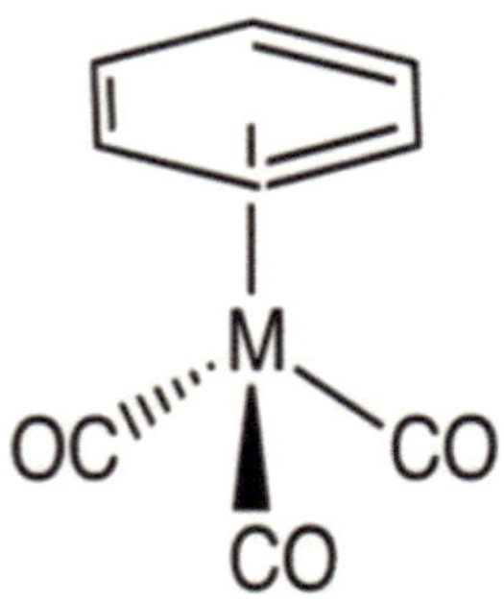

Figure 7. (η^6-C_6H_6) piano stool compound.

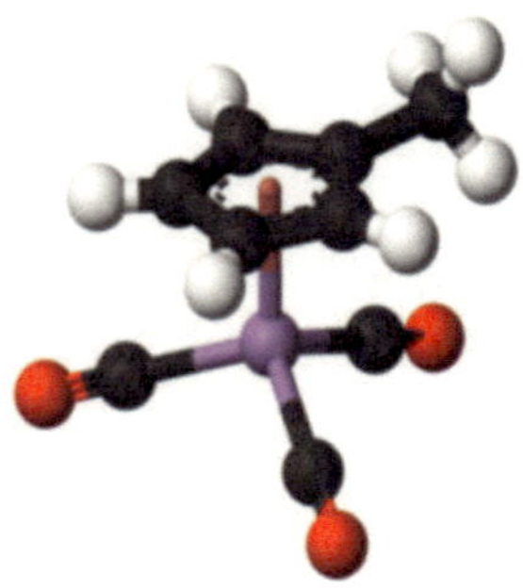

Figure 8. Ball-and-stick model of methylcyclopentadienyl manganese tricarbonyl, a "piano stool" compound.

http://dx.doi.org/10.5772/intechopen.68952

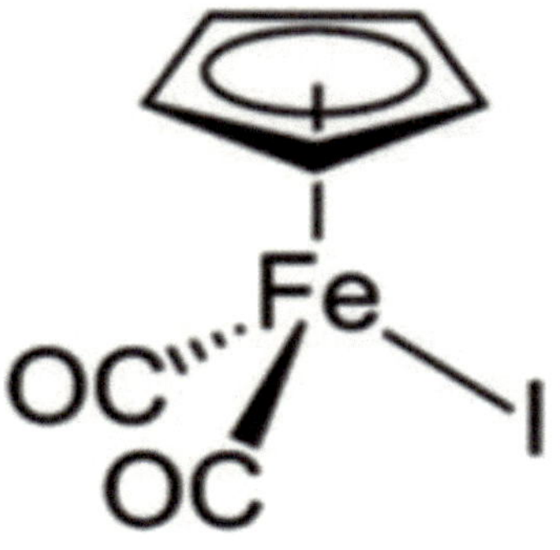

Figure 9. $CpFe(CO)_2I$ is an example of an unsymmetrical piano stool complex.

2.3. Bent metallocene

In organometallic chemistry, **bent metallocenes** are a subset of metallocenes. In bent metallocenes, the ring systems coordinated to the metal are not parallel but are tilted at an angle. A common example of a bent metallocene is Cp_2TiCl_2 [19, 20]. Titanocene dicarbonyl, an example of a bent metallocene, is given in **Figure 10**.

Bent metallocene dichlorides of molybdenum and tungsten are also prepared via indirect routes that involve redox at the metal centers.

Bent metallocenes have C_{2v} symmetry. The non-Cp ligands are arrayed in the wedge area. The bent metallocene is represented by the formula Cp_2ML_2. For bent metallocenes, the value of L-M-L angle affects the number of electrons. Molybdocene dichloride (Cp_2MoCl_2) has a Cl-Mo-Cl angle of 82°. In niobosen dichloride, this angle is slightly over 85.6°. For zirconocene dichloride, this angle is greater than 92.1°. This slope indicates that the boundary orbit is directed towards the MCl_2 plane. However, it expresses that the MCl_2 angle does not divide [21].

2.4. Multi-decker sandwich compounds

The dicationic triscyclopentadienyl dinickel complex $[Ni_2Cp_3](BF_4)_2$ was the first multidecker sandwich complex. Since this discovery, many multidecker sandwich compounds have been

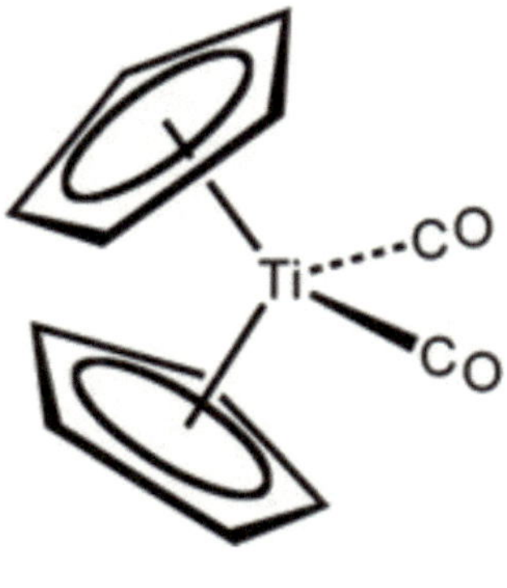

Figure 10. Titanocene dicarbonyl is an example of a bent metallocene.

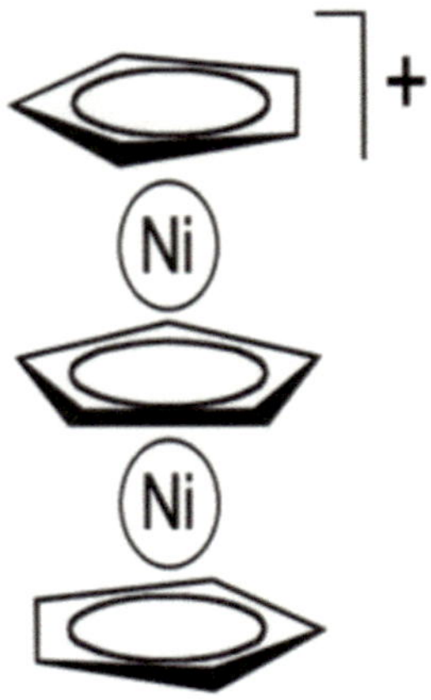

Figure 11. $[Ni_2Cp_3]^+$ is an example of triple-decker sandwich complex.

discovered. Especially, the triple deckers are the most important [22]. Different multi-decker sandwich complexes have been obtained by adding Cp^*Ru^+ to preformed sandwich complexes [23].

Triple-decker complexes are composed of three Cp anions and two metal cations in alternating order. The first triple-decker sandwich complex, $[Ni_2Cp_3]^+$, was reported in 1972. $[Ni_2Cp_3]^+$, an example of a triple-decker sandwich complex, is shown in **Figure 11**. Many examples have been reported subsequently, often with boron-containing rings [8].

3. Radical properties of metallocenes

3.1. Cyclopentadienyl radical and metal-substituted cyclopentadienyl radicals

EPR/spin trapping experiments were performed by direct irradiation of benzene solutions of the zirconocenes in the spectrometer cavity. Various spin traps have been used to determine the nature of **the cyclopentadienyl-type radical species**. These spin traps include *N-tert*-butyl-*α*-phenylnitrone (PBN), nitrosodurene (ND), and 5,5-dimethyl-1-pyrroline *N*-oxide (DMPO) [24, 25]. According to the NIEHS Spin Trap Database [24, 25], the hyperfine coupling constants (*hfcc*) of the spin adducts detected can be attributed to **the cyclopentadienyl type radicals** [26]. Therefore, the primary photochemical act can be described by **Figure 12**.

The presence of radical intermediates in the zirconocenes' photochemical reactions suggests a possible use as photoinitiators for radical polymerisation processes. In order to test the effectiveness of these organometallic photoinitiators, Similar EPR/spin trapping experiments were carried out in the presence of different alkenes such as 1-pentene, methyl methacrylate (MMA) and *tert*-butyl acrylate (*t*BA) and in different solvents such as benzene and dichloromethane. During irradiation of the zirconocene/alkene mixtures, the growth of the EPR signal of **a carbon-centred radical** was detected [26].

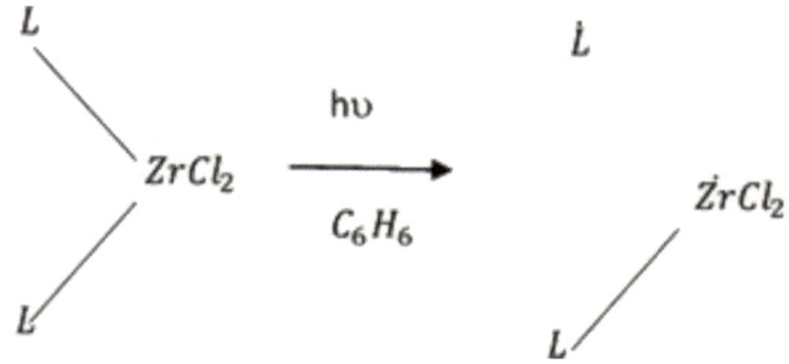

Figure 12. The formation of the cyclopentadienyl-type radical from the zirconocenes.

Due to the resonance state of **the cyclopentadienyl radical**, the bonds to the cyclopentadienyl ring are easily homolized. This situation is discussed in two respects. First, the photolysis is applied to a wide variety of cyclopentadienyl-metallic compounds by ultraviolet light. This effect leads **to the cyclopentadienyl radicals** and **the metal-centered radicals**. Some of these radicals have been characterized by EPR spectroscopy. Second, t-butoxyl radicals react with some other cyclopentadienyl-metallic compounds by removing hydrogen to give **the metal-substituted cyclopentadienyl radicals**.

The general characteristic of many cyclopentadienyl-metallic compounds is that they are not reactive. However, some are quite active. They also exhibit a stability profile for the cyclopentadienyl (5-anulene) ligand, where homolytic reactions are present. This causes the reactions to be homolytic. This has two important consequences. It causes the carbon-metal bond, which is the formation point of **the metal-centered radicals** and **the cyclopentadienyl radicals**, to be light-sensitive. Second, if bimolecular homolytic substitution does not readily occur at the metal center, hydrogen may be abstracted from the ring to give **a metal-substituted cyclopentadienyl radical** [27].

The photosensitivity work of the cyclopentadienyl-metal bond was first applied to tin(IV) derivatives. No significant EPR signal can be detected if simple alkyltin compounds are irradiated in solution with ultraviolet light in an EPR spectrometer cavity. Under the same conditions, cyclopentadienyltin(IV) compounds, η^1-$CpSnL_3$ (,L = Cp, alkyl, aryl, Cl, $MeCO_2$, etc.) give a strong sextet spectrum of cyclopentadienyl radicals [28]; this obscures the spectrum of the radical $\dot{S}nL_3$, but this radical can be identified by virtue of its characteristic reactions with reagents such as alkenes, alkyl halides, and 1,2-diones. For this reason, it is seen that it contains simple unimolecular homolysis. The pentahapto cyclopentadienyltin(II) compounds similarly show the spectrum of the $\dot{C}p$ radical, but an insoluble solid separates, and the fate of the tin moiety is unknown [29]. Similar studies have been carried out on cyclopentadienyl derivatives of other metals. The cyclopentadienyl derivatives of other metals, $CpML_n$, have been subjected to similar works. When M, lithium, mercury, tin(IV), tin(II), lead(IV), lead(II), titanium(IV), or zirconium(IV) are used as metals, M, the spectrum of the cyclopentadienyl radical is obtained. Cyclopentadienyl derivatives of beryllium, magnesium, boron, silicon, and germanium exhibit a very poor spectrum, but, in the presence of di-t-butyl peroxide, the compounds Cp_2Be, CpBeCl, Cp_2Mg, and $CpGeCl_3$ give the cyclopentadienyl radical under the effect of an S_H2 reaction in the metal center [30, 31].

When the metal carries both cyclopentadienyl and alkyl (R) groups, cleavage of the Cp—M and R—M bonds may be in competition. Irradiation of alkylcyclopentadienyltin(IV) compounds gives only the cyclopentadienyl radical but cyclopentadienyltriethyl-lead gives $\dot{C}p$ above −50°C and $\dot{E}t$ below −100°C, and both in between [28, 32]. Bis(cyclopentadienyl) zirconium dichloride give a rather weak $\dot{C}p$ spectrum, but Cp_2ZrMe_2 gives only the methyl radical [33].

Cyclopentadienylmethylberyllium reacts with t-butoxyl radicals to show predominantly the spectrum of the cyclopentadienyl radical below −60°C, but at −30°C the concentrations of cyclopentadienyl and methyl radicals are approximately equal [34]. However, with cyclopentadienylmethyl mercury at the normal concentration, the principal species which are observed are $\dot{C}p$ at −75°C, and $\dot{M}e$ at −130°C [35].

The EPR spectra have been studied with the parent cyclopentadienyl radical generated by several routes such as γ-irradiation on both crystalline [36, 37] and liquid cyclopentadiene [38], by pyrolysis of a molecular beam of ferrocene and azobenzene [39] and recently by hydrogen abstraction of cyclopentadiene with the *tert*-butoxy radical in solution [40, 41].

EPR spectroscopy has been used to study the photolysis of the compounds $(C_5H_5)_2ZrCl_2$, $(Me_5C_5)_2ZrCl_2$, $(C_5H_5)_2ZrMe_2$, $(C_5H_5)_2ZrMeCl$ and $(C_5Me_5)_2ZrMe_2$. The cyclopentadienyl radical is formed in the first two compounds, and the methyl radical is formed in the third and fourth compounds. But it can be said that both cyclopentadienyl and methyl radical exist in $(C_5Me_5)_2ZrMe_2$ compound [42].

The experimental and theoretical observations reported above are consistent with the notion that the reactive excited state in the bis(cyclopentadienyl)zirconium dichloride (Cp_2ZrCl_2), bis(pentamethyl-cyclopentadienyl)zirconium dichloride ($Cp_2^*ZrCl_2$) and bis(indenyl)zirconium dichloride (Ind_2ZrCl_2) complexes are of the L → Zr (L = Cp, Cp*, Ind) charge transfer type. Such the ligand-to-metal charge-transfer (LMCT) states would lead to formal reduction of the metal and oxidation of the ring system. Consistent with LMCT excitation, these excited states may dissociate, giving rise to free $\dot{L}$ and $L\dot{Z}rCl_2$ radicals. These results are consistent with the earlier reports of photochemistry of zirconium complexes parallel to titanium derivatives for $(C_5H_5)_2ZrCl_2$ [43].

The reaction shown in **Figure 12** is very consistent with the EPR results. In this reaction, the photo-induced cleavage of one of the L—Zr bonds as the primary excited state reaction is generally believed to be the case [44].

3.2. Anion and cation radicals formed in metallocenes

The redox, photophysical, and photochemical properties of the homologous bent metallocenes of group 4 transition metals are emphasized. On the systematic variation of the definition of ligands of metal ions (Ti, Zr, or Hf), auxiliary π- and monodentate σ- (Cl, Me) ligands, a comparative analysis of electron transfer induced transformations and ligand-to-metal charge transfer excited states was carried out for bent metalocene complexes. Linear correlations between optical and redox HOMO-LUMO electron transitions are found for such organometallic π-complexes. It is proposed that the combination of spectroscopic and electrochemical techniques provide important diagnostics to determine "ionization potential" and "electron affinity" in the solution and the energy gap in metalocene complexes [45].

For the synthesis of dithiolane complexes [Cp(2)M(S(2)C(2)(H)R)] (M = Mo or W; R = phenyl, pyridin-2-yl, pyridin-3-yl, pyridin-4-yl or quinoxalin-2-yl) and [Cp(2)Mo(S(2)C(2)(Me)(pyridin-2-yl)] compounds were prepared. These compounds are electrochemically subjected to one-electron reduction and one-electron oxidation process. For a Mo compound, each redox exchange occurs at a more positive potential than the value for a W compound. Monocations of both compounds were produced by chemical and electrochemical oxidation. For [Cp(2)Mo(S(2)C(2)(H)R)](+)/[Cp(2)Mo(S(2)C(2)(H)R)] (R = Ph or pyridin-3-yl) redox pairs, the changes in Mo—S, SC and CC bond lengths of the {MoSCCS} moiety are consistent with the oxidation process involving an electron loss from the π-orbital in Mo—S and C—S antibonding and C—C bonding. When the EPR spectrum of each Mo cation is examined, it is understood that the unpaired electron is weakly bound to the ditiolan proton. According to the results of DFT calculations, the unpaired electrons in the monoanions are localized on the metal, more than mono-cations. Furthermore, according to the EPR spectrum, the hyperfine structure splits of mono anions containing Mo are larger than those of mono cations. The reduction of [Cp(2)W(S(2)C(2)(H)(quinoxalin-2-yl)] takes place at a more positive potential than expected for Mo. The EPR spectrum of the mono-anion is typical for an organic radical. DFT calculations show that these properties are due to the addition of a the electron to quinoxalin-2-yl π-orbital [46].

The cyclic voltammetry of cobaltocene and nickelocene revealed five redox states, from dication to dianion. In situations where it is not possible to work with conventional electrodes at traditional temperatures, operation potential can be extended using low-temperature solvents (SO_2, THF) and ultramicroelectrode techniques. The $Cp_2M^{0/-}$, $Cp_2M^{-/2-}$, and $Cp_2M^{+/2+}$ redox reactions of the previously known $Cp_2M^{0/+}$ redox couples are shown [47].

The classical metalocenes, ferrocene, and ruthenocene are easily soluble in the pure ethyl 2-cyanoacrylate (CA) monomer. The electronic spectra of the resulting solutions show a near-ultraviolet absorption band as a result of the charge-transfer transition to the solvent (metallocene → CA). The one-electron oxidation of the metallocene occurs when this band is exposed to irradiation. The anionic polymerization of the electrophilic monomer begins with the addition of the latter species to CA [48].

A large series of studies have been devoted to metallocenes, starting with the simplest representatives of compounds of this class-ferrocene [49] and its derivatives [50–55]. Comparison of the results of the calculation by the Hückel molecular orbital method with the observed hyperfine structure of EPR spectra [51] yields a quantitative distribution of the spin density in the radical-anion. Overall, the ferrocenyl group is destabilizing compared with the phenyl group. In the presence of both groups in one molecule, the degree of delocalization of the spin density is greater in the direction of the benzene ring [52].

Stable paramagnetic semiquinones of the metallocene series are formed on oxidation of ketones containing the a methylene group [53, 54].

The ease of the one-electron oxidation of metallocene derivatives is confirmed by the formation of paramagnetic salts, where the anionic component may consist of the radical-anion derived from the organic one-electron acceptor tetracyanoquinodimethane (TCNQ) [56]. Such salts are in essence organometallic analogues of charge transfer complexes [57].

Ferrocene and its derivatives turn into cation radicals by the reversible one-electron oxidation. These radicals are called "ferricenium" cations. The iron atom is the center of the cation radical and localized. Conversely, the hole transfers via conjugated systems were proven for bis(ferrocenyl) ethylene cation radical [58] and thebis (fulvaleneiron) cation radical [59]. The unpaired electron is delocalise on both metallocene residues in the bis (fulvaleneiron) cation radical, and Extended Huckel MO calculations [60] support this situation. Alternatively, very rapid intra-ionic intervalence electron transfer can occur between the formal Fe(II) and Fe(III) atoms. In general, the ethylene bonds in the organic cation radicals are weak and the barrier to rotation is significantly reduced relative to the neutral ethylene derivative. This property of the ethylene bond in the cation radicals has been used to investigate many reaction mechanisms [61]. The cation radicals of the ferrocenyl ethylene are not subject to the cis-to-trans isomerization process. The calculations show that the cation radical center is only in the iron atom, not in the ethylene bond [62]. Therefore, one-electron oxidation of ferrocenyl ethylene occurs at the iron atom. For this reason, the stable enol bound to a ferrocene redox center gives a cation radical through an electron oxidation. This species is characterized as a ferricenic salt rather than an enol cation radical [63].

The cationic forms of the metalocene (the metallocinium cations) provide interesting possibilities by forming charge-transfer complexes with various biological macromolecules. Although the suitability of the metallocene compounds has not been discussed yet, Szent-Györgyi emphasized that the charge-transfer interactions of metallocene compounds are of interest to biological problems [64].

The basic coenzymes are all conjugate compounds: DPN, TPN, FAD, FMN, quinones, folic acids, pyridoxal phosphate, etc. The all of the significant coenzymes are all conjugated compounds: DPN, TPN, FAD, FMN, quinones, folic acids, pyridoxal phosphate, etc. Likewise, steroids and a large numbers of pharmacological agents contain suitable conjugate moieties for the charge transfer interactions. The electronic delocalization is thought to be one of their main features [65]. It seemed of interest, therefore, to examine the possible electron-donor properties of some biologically important molecules, particularly of conjugated systems likely to be involved in electron-transfer phenomena in relation to the acceptor characteristics of the metallocenes. The capacity of metallocinium cations to accept π-electrons might, in fact, be employed toward this end. They could serve as probes of the electron-donating characteristics of conjugated biological systems. The ferricinium, the bis-benzene-chromium, and the cobalticinium cations apparently display the required physicochemical properties. All these compounds have aromatic properties. The ferricinium, the bis-benzene-chromium, and the cobalticinium cations contain two parallel conjugated rings, which are either five- or six-membered. They are readily soluble in water. Aqueous solutions of the ferricinium salts [66] and bis-benzene-chromium salts [67] are reasonably stable in the dark.

The cobalticinium salts are exceptionally stable in water [68]. The simple application of a metallic cation between the two aromatic rings has three important consequences for the investigation of biological systems:

(1) The aromatic system becomes more soluble in water.

(2) It becomes a good electron acceptor.

(3) The tightly bound metal ion can be identified by optical, EPR and NMR spectra and is an excellent label of the aromatic system.

The solubility in water of the metallocinium cations has a distinct advantage. Most of the conventional π-acceptors can only be examined in less polar solvents such as chloroform or dimethylformamide [69].

The formation of stable η^5-complexes between fullerenes and transition metals is highly improbable because the conjugated system is rather strongly delocalized, and the polarization of the atoms of the five-member face is weak. There is an opinion that the fullerene structure will change by the addition of R groups, hydrogen (H), methyl (Me), or phenyl (Ph) to the α-position of the same five-membered face, and thus the conjugate system will deteriorate.

The cyclopentadienyl ion obtained from the fullerene plays an important role in the formation of complexes with transition metals [70]. The molecular structure of ferrocene/C_{70}-fullerene hybrid [71] is given in **Figure 13**. There are two obstacles to producing the cyclopentadienyl

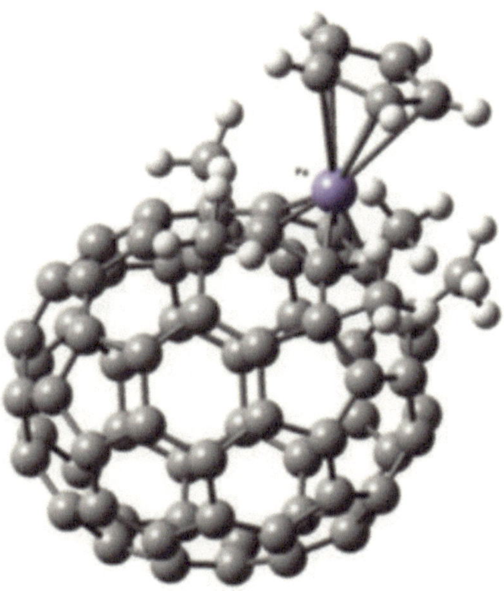

Figure 13. Molecular structure of ferrocene/C_{70}-fullerene hybrid.

ion: first, it is difficult to form a barrier around the cyclopentadienyl ring. And the second is the difficulty in ionizing to produce the cyclopentadienyl group.

The same idea, that is, double-decker ferrocene complexes, has also been examined by the Nakamura group [72, 73]. They studied both theoretical and experimental arguments and found that the addition of a second ferrocene fraction leads to strong instability and can produce very short-lived radical ion pairs [74]. The complex $2\eta^5$-π-$(CpFe)_2$-$C_{60}H_{10}$ [75], double-decker ferrocene complex, is shown in **Figure 14**.

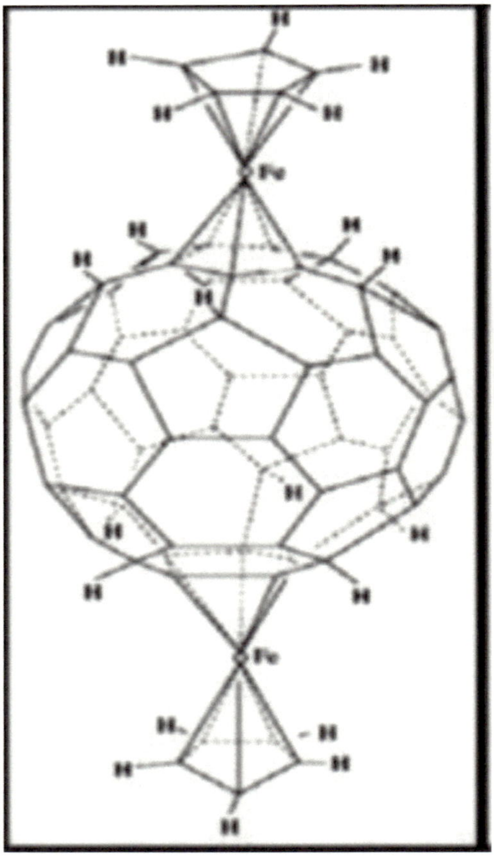

Figure 14. Complex $2\eta^5$-π-$(CpFe)_2$-$C_{60}H_{10}$.

4. Conclusion

In metallocenes, the radicals can occur in a variety of ways. Photolysis (ultraviolet light), radiolysis (gamma-irradiation), spin trapping and photochemical reactions can be created radical. It has been observed that a wide variety of radicals are formed by these methods. The ionic radicals are formed through charge-transfer interactions.

Metallocenes can form a wide variety of radicals. However, the two main types of radical groups gain weight. These are the cyclopentadienyl radicals and the ionic radicals. The cyclopentadienyl radicals are distinguished as the cyclopentadienyl radical and the metal-centered cyclopentadienyl radical.

Other radicals that can occur outside of these radicals can be expressed as follows: the carbon-centred radical, the metal-centred radical, various anion and cation radicals of metallocenes, dianion and dication radicals, and semiquinone radicals of metallocenes.

Author details

Betul Caliskan

Address all correspondence to: bcaliska@gmail.com

Department of Physics, Faculty of Arts and Science, Pamukkale University, Kinikli, Denizli, Turkey

References

[1] Kealy TJ, Pauson PL. A new type of organo-iron compound. Nature. 1951;**168**(4285):1039-1040. DOI: 10.1038/1681039b0

[2] Miller SA, Tebboth JA, Tremaine JF. Dicyclopentadienyliron. Journal of the Chemical Society. 1952:632-635. DOI: 10.1039/JR9520000632

[3] Wilkinson G, Rosenblum M, Whiting MC, Woodward RB. The structure of iron bis-cyclopentadienyl. Journal of the American Chemical Society. 1952;**74**(8):2125-2126. DOI: 10.1021/ja01128a527

[4] Fischer EO, Pfab W. Cyclopentadien-metallkomplexe, ein neuer typ metallorganischer verbindungen. Zeitschrift für Naturforschung B. 1952;**7**(7):377-379. DOI: 10.1515/znb-1952-0701

[5] Chirik PJ. Group 4 transition metal sandwich complexes: Still fresh after almost 60 years. Organometallics. 2010;**29**(7):1500-1517. DOI: 10.1021/om100016p

[6] Brennan JG, Andersen RA, Zalkin A. Chemistry of trivalent uranium metallocenes: Electron-transfer reactions. Synthesis and characterization of [(MeC5H4)3U]2E (E=S, Se, Te) and the crystal structures of hexakis(methylcyclopentadienyl)sulfidodiuranium and tris(methylcyclopentadienyl)(triphenylphosphine oxide)uranium. Inorganic Chemistry. 1986;**25**(11):1761-1765. DOI: 10.1021/ic00231a008

[7] Long NJ. Metallocenes: Introduction to Sandwich Complexes. London: Wiley-Blackwell, 1998

[8] Grimes RN. Boron clusters come of age. Journal of Chemical Education. 2004;**81**(5):657-672. DOI: 10.1021/ed081p657

[9] Dunitz JD, Orgel LE, Rich A. The crystal structure of ferrocene. Acta Crystallographica. 1956;**9**(4):373-375. DOI: 10.1107/S0365110X56001091

[10] Miessler GL, Tarr DA. Inorganic Chemistry. 3rd ed. Upper Saddle River, NJ: Pearson Prentice Hall; 2004

[11] Urnius E, Brennessel WW, Cramer CJ, Ellis JE, Schleyer P von R. A carbon-free sandwich complex [(P5)2Ti]2-. Science. 2002;**295**(5556):832-834. DOI: 10.1126/science.1067325

[12] Elschenbroich C. Organometallics. 3rd ed. Weinheim: Wiley-VCH; 2006

[13] Natta G, Ercoli R, Calderazzo F. (η-C6H6)Cr(CO)3. Chimica e Industria. 1958;**40**(12):1003-1007

[14] Fischer EO, Öfele K, Essler H, Fröhlich W, Mortensen JP, Semmlinger W. Über Aromatenkomplexe von metallen, XXIV. Über gemischte tricarbonylkomplexe des chroms, molybdäns und wolframs mit benzol und seinen derivaten. Chemische Berichte. 1958;**91**(12):2763-2772. DOI: 10.1002/cber.19580911231

[15] Fischer EO, Öfele K. Über aromatenkomplexe von metallen, XIII benzol-chrom-tricarbonyl. Chemische Berichte. 1957;**90**(11):2532-2535. DOI: 10.1002/cber.19570901117

[16] Corradini P, Allegra G. X-ray determination of the structure of tricarbonylchromium-benzene. Journal of the American Chemical Society. 1959;**81**(9):2271-2272. DOI: 10.1021/ja01518a065

[17] Begley MJ, Puntambekar SG, Wirght AH. A di-iron-anthracene complex via ultrasonics. Journal of the American Chemical Society. Chemical Communications. 1987;(16):1251-1252. DOI: 10.1039/C39870001251

[18] Begley MJ, Puntambekar SG, Wright AH. Synthesis and reactivity of a new class of half-sandwich arene-iron complex: Structure of $[C_6H_3Me_3Fe(C_3H_5)(CO)]PF_6$. Journal of Organometallic Chemistry. 1989;**362**(1-2):C11–C14. DOI: 10.1016/0022-328X(89)85301-X

[19] Green JC. Bent metallocenes revisited. Chemical Society Reviews. 1998;**27**(4):263-272. DOI: 10.1039/A827263Z

[20] Erker G, Kehr G, Fröhlich R. Group 4 bent metallocenes and functional groups—Finding convenient pathways in a difficult terrain. Coordination Chemistry Reviews. 2006;**250**(1-2):36-46. DOI: 10.1016/j.ccr.2005.04.006

[21] Prout K, Cameron TS, Forder RA, Critchley SR, Denton B, Rees GV. The crystal and molecular structures of bent bis-π-cyclopentadienyl-metal complexes: (*a*) bis-π-cyclopentadienyldibromorhenium(V) tetrafluoroborate, (*b*) bis-π-cyclopentadienyldi chloromolybdenum(IV),(*c*)bis-π-yclopentadienylhydroxomethylaminomolybdenum (IV) hexafluorophosphate, (*d*) bis-π-cyclopentadienylethylchloromolybdenum(IV), (*e*) bis-π-cyclopentadienyldichloroniobium(IV), (*f*) bis-π-cyclopentadienyldichloromolyb denum(V) tetrafluoroborate, (*g*) μ-oxo-bis[bis-π-cyclopentadienylchloroniobium(IV)]

tetrafluoroborate, (*h*) bis-π-cyclopentadienyldichlorozirconium. Acta Crystallographica. 1974;**B30**:2290-2304. DOI:10.1107/S0567740874007011

[22] Beck V, O'Hare D. Triple-decker transition metal complexes bridged by a single carbocyclic ring. Journal of Organometallic Chemistry. 2004;**689**(24):3920-3938. DOI: 10.1016/j.jorganchem.2004.06.011

[23] Fagan PJ, Ward MD, Calabrese JC. Molecular engineering of solid-state materials: Organometallic building blocks. Journal of the American Chemical Society. 1989;**111**(5):1698-1719. DOI: 10.1021/ja00187a024

[24] Sostero S, Rehorek D, Polo E, Traverso O. Spin trapping and some reactions of ruthenium centred radicals. Inorganica Chimica Acta. 1993;**209**(2):171-176. DOI: 10.1016/S0020-1693(00)85138-5

[25] di Martino S, Sostero S, Traverso O, Rehorek D, Kemp TJ. Spin trapping of radicals formed during the photolysis of the dimeric iron carbonyl complexes $[\eta^5\text{-}C_5H_5)Fe(CO)_2]_2$ and $[(\eta^5\text{-}C_5Me_5)Fe(CO)_2]_2$. Inorganica Chimica Acta. 1990;**176**(1):107-112. DOI: 10.1016/S0020-1693(00)85100-2

[26] Polo E, Barbieri A, Sostero S, Green MLH. Zirconocenes as photoinitiators for free-radical polymerisation of acrylates. European Journal of Inorganic Chemistry. 2002;**2**:405-409. DOI: 10.1002/1099-0682(20022)2002:2"405::AID-EJIC405"3.0.CO;2-G

[27] Davies AG. Photolytic reactions of cyclopentadienylmetallic compounds. Pure and Applied Chemistry. 1982;**54**(1):23-28. DOI: 10.1351/pac198254010023

[28] Barker PJ, Davies AG, Tse M-W. The photolysis of cyclopentadienyl compounds of tin and mercury. Electron spin resonance spectra and electronic configuration of the cyclopentadienyl, deuteriocyclopentadienyl, and alkylcyclopentadienyl radicals. Journal of the Chemical Society, Perkin Transactions 2. 1980;(6):941-948. DOI: 10.1039/P29800000941

[29] Kira M, Watanabe M, Sakurai H. Chemistry of organosilicon compounds. 131. Substituent effects by deuterium and alkyl groups and carbon-13 hyperfine coupling constants of cyclopentadienyl radicals as studied by electron spin resonance. Journal of the American Chemical Society. 1980;**102**(16):5202-5207. DOI: 10.1021/ja00536a014

[30] Kira M, Watanabe M, Sakurai H. Chemistry of organosilicon compounds. 100. Substituent effects on the electron spin resonance spectra of cyclopentadienyl radicals. Removal of degeneracy by organosilyl groups. Journal of the American Chemical Society. 1977;**99**(24):7780-7785. DOI: 10.1021/ja00466a006

[31] Barker PJ, Davies AG, Henriquez R, Nedelec J-Y. An electron spin resonance study of the effect of group IV substituents in cyclopentadienyl radicals. Journal of the Chemical Society, Perkin Transactions 2. 1982;**6**:745-750. DOI: 10.1039/P29820000745

[32] Davies AG, Hawari JA-A, Gaffney C, Harrison PG. An electron spin resonance study of the photolysis of some cyclopentadienyl-lead compounds: the comparison with tin.

Journal of the Chemical Society, Perkin Transactions 2. 1982;**5**:631-635. DOI: 10.1039/P29820000631

[33] Brindley PB, Davies AG, Hawari JA-A. An ESR study of the photolysis of dicyclopentadienyltitanium dichloride. Journal of Organometallic Chemistry. 1983;**250**(1):247-256. DOI: 10.1016/0022-328X(83)85055-4

[34] Davies AG, Lusztyk E, Lusztyk J. An electron spin resonance study of the methylcyclopentadienyl radicals, $Me_nH_{5-n}C_5\cdot$(n= 1-5). Journal of the Chemical Society, Perkin Transactions 2. 1982;**6**:729-736. DOI: 10.1039/P29820000729

[35] Davies AG, Lusztyk J. The pentamethylcyclopentadienyl radical: generation, electron spin resonance spectrum, and reaction kinetics. Journal of the Chemical Society, Chemical Communications. 1980;**12**:554-555. DOI: 10.1039/C39800000554

[36] Ohnishi S-I, Sugimoto S-I, Nitta I. Photoinduced change of organic free radicals as studied by ESR. I. Conversion of allyl-type to alkyl-type radicals in irradiated polyethylene. The Journal of Chemical Physics. 1963;**39**(10):2647-2653. DOI: 10.1063/1.1734077

[37] Liebling GR, McConnell HM. Study of molecular orbital degeneracy in C_5H_5. The Journal of Chemical Physics. 1965;**42**(11):3931-3934. DOI: http://dx.doi.org/10.1063/1.1695862

[38] Fessenden RW, Ogawa S. On variations in the hyperfine splitting of benzene negative ion and on the value of Q. Journal of the American Chemical Society. 1964;**86**(17):3591-3592. DOI: 10.1021/ja01071a062

[39] Zandstra PJ. Electron spin resonance of cyclopentadienyl. The Journal of Chemical Physics. 1964;**40**(2):612-612. DOI: 10.1063/1.1725177

[40] Krusic PJ, Kochi JK. Electron spin resonance of aliphatic hydrocarbon radicals in solution. Journal of the American Chemical Society. 1968;**90**(25):7155-7157. DOI: 10.1021/ja01027a065

[41] Kawamura T, Kochi JK. The structure and conformation of stannyl and related radical adducts to cyclopentadiene by ESR. Journal of Organometallic Chemistry. 1973;**47**(1):79-88. DOI: 10.1016/S0022-328X(00)92842-0

[42] Atkinson JM, Brindley PB, Davies AG, Hawari JA-A. An ESR study of the photolysis of cyclopentadienylzirconium compounds. Journal of Organometallic Chemistry. 1984;**264**(1-2):253-261. DOI: 10.1016/0022-328X(84)85151-7

[43] Tsai Z-T, Brubaker Jr., CH. Photolysis of titanocene dichloride. Journal of Organometallic Chemistry. 1979;**166**(2):199-210. DOI: 10.1016/S0022-328X(00)91634-6

[44] Barbieri A, Droghetti A, Sostero S, Traverso O. Photochemistry of *ansa*-zirconocenes: ethylene-bis(1-indenyl)- and ethylene-bis(4, 7-dimethyl-1-indenyl) zirconium dichlorides. Journal of Photochemistry and Photobiology A: Chemistry. 1999;**129**(3):137-142. DOI: 10.1016/S1010-6030(99)00167-7

[45] Loukova GV, Strelets VV. A review on molecular electrochemistry of metallocene dichloride and dimethyl complexes of group 4 metals: Redox properties and relation with

optical ligand-to-metal charge transfer transitions. Collection of Czechoslovak Chemical Communications. 2001;**66**:185-206. DOI: 10.1135/cccc20010185

[46] Whalley AL, Blake, AJ, Collison D, Davies ES, Disley HJ, Helliwell M, Mabbs FE, McMaster J, Wilson C, Garner CD. Synthesis, structure and redox properties of bis(cyclopentadienyl) dithiolene complexes of molybdenum and tungsten. Dalton Transactions. 2011;**40**:10457-10472. DOI: 10.1039/C1DT10663E

[47] Bard AJ, Garcia E, Kukharenko S, Strelets VV. Electrochemistry of metallocenes at very negative and very positive potentials. Electrogeneration of 17-electron Cp2Co2+, 21-electron Cp2Co2-, and 22-electron Cp2Ni2- species. Inorganic Chemistry. 1993;**32**(16):3528-3531. DOI: 10.1021/ic00068a024

[48] Sanderson CT, Palmer BJ, Morgan A, Murphy M, Dluhy RA, Mize T, Amster IJ, Kutal C. Classical metallocenes as photoinitiators for the anionic polymerization of an alkyl 2-cyanoacrylate. Macromolecules. 2002; **35**(26):9648-9652. DOI: 10.1021/ma0212238

[49] McDonnell JJ, Cafen G, Michealson R. Paramagnetic metallocenes: Interannular electron transfer. Tetrahedron Letters 1969;**10**(49):4251-4254. DOI: 10.1016/S0040-4039(01)88667-6

[50] Danen WC, West CT. Electron spin resonance study of radical anions derived from substituted ferrocenes. Tetrahedron Letters, 1970;**11**(3):219-222. DOI: 10.1016/0040-4039(70)80030-2

[51] Elschenbroich C, Cais M. Organometallic studies XXIX. Electron spin resonance of ferrocene-substituted radical anions. Journal of Organometallic Chemistry. 1969;**18**(1):135-143. DOI: 10.1016/S0022-328X(00)80241-7

[52] Bigam G, Hooz J, Linke S, McClung RED, Mosher MW, Tanner DD. An electron spin resonance examination of ferrocenylalkyl- and aryl ketyls. Canadian Journal of Chemistry. 1972;**50**(12):1825-1830. DOI: 10.1139/v72-293

[53] McDonnell JJ, Pochopien DJ. Paramagnetic metallocenes. Oxidation of ferrocenylketnes. The Journal of Organic Chemistry. 1971;**36**(15):2092-2098. DOI: 10.1021/jo00814a014

[54] McDonnell JJ, Pochopien DJ. Ferrocenebenzosemiquinones. The Journal of Organic Chemistry. 1972;**37**(25):4064-4067. DOI: 10.1021/jo00798a018

[55] McDonnell JJ. Oxidation of ferrocenyl ketones. Tetrahedron Letters. 1969;**10**(25):2039-2042. DOI: 10.1016/S0040-4039(01)88080-1

[56] Reis AH Jr, Preston LD, Williams JM, Peterson SW, Candela, GA, Swartzendruber LJ, Miller JS. Crystal and molecular structure of the paramagnetic 1:1 decamethylferrocenium 7, 7, 8, 8-tetracyano-p-quinodimethanide dimer salt: {[Fe(C5Me5)2]+.cntdot.}2(TCNQ)22-. Journal of the American Chemical Society. 1979;**101**(10):2756-2758. DOI: 10.1021/ja00504a058

[57] Milaeva ER, Rubezhov AZ, Prokof'ev AI, Okhlobystin OYu. The unpaired electron in transition metal complexes. Russian Chemical Reviews. 1982;**51**(10):1638-1673

[58] Delgado-Pena F, Talham DR, Cowan DO. Near-IR spectroscopic studies of mixed-valence di-, tri-, and tetraferrocene derivatives. Journal of Organometallic Chemistry. 1983;**253**(3):C43-C46. DOI: 10.1016/S0022-328X(00)99241-6

[59] Kirchner RF, Loew GH, Mueller-Westerhoff UT. Theoretical study of the electromagnetic properties of bis(fulvalene)diiron in its three oxidation states. Inorganic Chemistry. 1976;**15**(11):2665-2670. DOI: 10.1021/ic50165a020

[60] LeVanda C, Bechgaard K, Cowan DO, Mueller-Westerhoff UT, Eilbracht P, Candela GA, Collins RL. Bis(fulvalene)diiron, its mono- and dications. Intramolecular exchange interactions in a rigid system. Journal of the American Chemical Society. 1976;**98**(11):3181-3187. DOI: 10.1021/ja00427a021

[61] Schmittel M, Langels A. Enol radical cations in solution. Part 12. Synthesis and electrochemical investigations of a stable enol linked to a ferrocene redox centre. Journal of the Chemical Society, Perkin Transactions 2. 1998;(3):565-572. DOI: 10.1039/A707743B

[62] Todres ZV. Tetrahedron report number 224: Ethylenic compounds as probes for the study of donor-acceptor interaction. Tetrahedron 1987;**43**(17):3839-3861. DOI: 10.1016/S0040-4020(01)81667-7

[63] Todres ZV. *Cis/trans* conversion of potassium derivatives of 2- and 4-nitrostilbenes. Journal of Organometallic Chemistry. 1992;**441**(3):349-354. DOI: 10.1016/0022-328X(92)80165-T

[64] Szent-Györgyi A. Introduction to a Submolecular Biology. Academic Press; New York, 1960. pp. 67-74. DOI: 10.1016/B978-0-12-395612-5.50010-5

[65] Szent-Györgyi A. Horizons in Biochemistry. Kasha M, Pullman B, editors. Academic Press; New York, NY. 1962. pp. 564-580

[66] Pauson PL. Organometallic Chemistry. Zeiss H, editor. Reinhold Publishing Corporation; New York, NY. 1960

[67] Zeiss HH, Wheatley PJ, Winkler HJS. Benzenoid Metal Complexes: Structural determinations and chemistry. The Ronald Press Company; New York, NY. 1966

[68] Cotton FA. Progress in Inorganic Chemistry. Chapter-1. In: Wilkinson G, Cotton FA, editors. Cyclopentadienyl and Arene Metal Compounds). 1959. pp. 1-124. DOI: 10.1002/9780470166024.ch1

[69] Briegleb G. Elektronen Donator-Acceptor Komplexe. Berlin, Germany: Springer-Verlag; 1961

[70] Chistyakov AL, Stankevich IV. Cyclopentadienyl type η^5-π-complexes of C_{60} fullerene derivatives with indium and thallium: Simulation of molecular and electronic structure by the MNDO/PM3 method. Russian Chemical Bulletin. 1997;**46**(11):1832-1837. DOI: 10.1007/BF02503767

[71] Kaji T, Shimada T, Inoue H, Kuninobu Y, Matsuo Y, Nakamura E, Saiki K. Molecular orientation and electronic structure of epitaxial bucky ferrocene ($Fe(C_{60}(CH_3)_5)C_5H_5)$) thin films. The Journal of Physical Chemistry B. 2004;**108**(28):9914-9918. DOI: 10.1021/jp037668v

[72] Matsuo Y, Tahara K, Nakamura E. Synthesis and electrochemistry of double-decker buckyferrocenes. Journal of the American Chemical Society. 2006;**128**(22):7154-7155. DOI: 10.1021/ja061175x

[73] Marczak R, Wielopolski M, Gayathri SS, Guldi DM, Matsuo Y, Matsuo K, Tahara K, Nakamura E. Uniquely shaped double-decker buckyferrocenes—Distinct electron donor–acceptor interactions. Journal of the American Chemical Society. 2008;**130**(48):16207-16215. DOI: 10.1021/ja8013902

[74] Soto D, Salcedo R. Coordination modes and different hapticities for fullerene organometallic complexes. Molecules. 2012;**17**(6):7151-7168. DOI: 10.3390/molecules17067151

[75] Matsuo Y, Iwashita A, Nakamura E. Synthesis and derivatization of iridium(i) and iridium(iii) pentamethyl[60]fullerene complexes. Organometallics. 2005;**24**(1):89-95. DOI: 10.1021/om049333l

Concerning Organometallic Compounds in Environment: Occurrence, Fate, and Impact

Kovacs Melinda Haydee and Kovacs Emoke Dalma

Additional information is available at the end of the chapter

http://dx.doi.org/10.5772/67755

Abstract

Organometallic compounds can be found in our surrounding environmental compartments either because of human extensive activities or their existence as natural products in the environment. Since organometallic species of trace metals were found often more worrying than their parent compounds, intensive research on their properties, pathways of transformation in different environmental compartment as well as their fate and interactions between different environmental compartments (under different external and internal conditions), and not finally their end-up and disposal, has become a requirement from many public health and environmental protection agencies.

Keywords: organomercury, organotin, organolead, environment

1. Introduction

In environment, most of organometallic compounds were found as persistent that do not easily decompose but are easily concentrated and are highly toxic, often more toxic than their elemental form Refs. [1, 2]. These compounds could be characterized as they have a metal or metalloid-carbon bond. Usually the bonds in them are covalent and are reached between soft acid metals and soft ligands. In environment, their fate and pathways occur through interactions with other chemicals and biota from the environmental compartment, their cycling and pathways between multiple environmental compartments and phases are physically, chemically, and biologically mediated [3].

Concerns considering organometallic compounds are accentuated also by that once they accumulate in different environmental compartments can inhibit their functioning and consequently affect the ecosystem, plants, and other living organisms (micro and macro), and also contaminate the food chain (including that of humans) [4–6].

As natural components of the Earth's crust and resultant of biogeochemical reactions, organometallic compounds are generally present at low concentrations in natural environmental compartments, as soil or water, but extensive anthropogenic activities over the past 50 years (industrial, mining, agricultural, and urban-extension activities) have greatly increased their inputs in different compartments of our surrounding environment; thus,their presence is becoming a severe problem at worldwide level [7–10].

Among all compartments, soil plays an important role in the distribution and fate of organometallic compounds, since often it serves as a major reservoir and sink of these pollutants due to its large absorption capacity [11, 12]. In the terrestrial and aquatic environment, metals occur in both organic and inorganic forms, including elemental forms, salts, and organometallic compounds. In most cases, the mineral form of metals is insoluble thus rendering these species rather unavailable for transport or plant uptake in the short term [5].

Inorganic species are adsorbed onto soil organic matter and/or metal oxides which can be subjected to biogeochemical processes (e.g., reduction, oxidation, methylation, alkylation, and biomethylation) resulting in highly mobile organic species that further have the ability to form water-soluble complexes in living organism body tissues thus increasing the potential for uptake and accumulation by organisms [13, 14]. Moreover, bond formation between methyl, ethyl, or alkyl groups and metals or metalloids cause changes in their physical properties as solubility or volatility, properties that could significantly affix their fate, pathways, and life cycle both in the original environmental compartment and between interconnected environmental compartments. The rates of all these processes depend greatly on the local conditions, which exist in the relevant ecosystem and the microbial activity, as the pollutant's leaching rate strongly depends on its specific geochemical properties [15, 16].

2. Concerning organometallic compounds occurrence in environment from anthropogenic sources

The compounds considered in this work are those having environmental implications and are susceptible to threat biota and human health. Thus, we limited this chapter to organic forms of mercury, lead, and tin. These compounds could occur in environment either naturally or deposited as an industrial pollutant.

2.1. Mercury and its organic derivatives

Mercury (Hg) is a nonessential and extremely toxic trace element that poses global environmental and human health risks [5, 13]. Its biogeochemical cycle was perturbed during the last centuries by anthropogenic inputs. The organic forms of mercury compounds have been used in chlor-alkali plants and coal power station industry, and also in other anthropogenic activities, such as catalysts, fungicides, herbicides, disinfectants, and pigments [5, 12, 13]. Emissions and inputs from those mentioned industrial processes as well as from the combustion of fossil fuels and waste disposal finally resulted in severe environmental contamination [17, 18].

As mentioned in many studies, mercury and its related compounds are considered health hazards, but their toxicity depends strongly on their chemical forms, those organic forms of mercury, such as methylmercury or dialkyl mercury, are considered more toxic than inorganic salts of mercury [19, 20].

(1) Generally, metal speciation marks both the fate as well as the toxicity of metals in environmental compartments, that speciation adverts to the occurrence of the different variety of chemical forms of a specific metal in the environment. Such forms of metals could be free ions, complexes (dissolved in solution or sorbed on solid surfaces) or as forms that have been coprecipitated in major metal solids or which occur in their own solids [3]. Considering scientific reports regarding mercury species toxicological effects, it becomes necessary to speciate mercury [21]. Lindqvist et al. [22] categorized the mercury species compounds into three categories: (i) volatile species (Hg); (ii) reactive species (Hg^{2+}, HgO on aerosol particles, Hg^{2+} complexes with OH, Cl, Br, and organic acids); and (iii) nonreactive species (CH_3Hg^+ and other organomercurial moieties, $Hg(CN)_2$, HgS and Hg^{2+} bond to sulfur in fragments of humic matter) [20]. The speciation of mercury and its related compounds affects, besides the degree of toxicity, also its properties (e.g., volatilization, photolysis, sorption, atmospheric deposition, acid/base equilibrium, diffusivity, microbial transformation degree, and pattern) that characterize their fate and pathways in the environment [3]. The organic species of mercury that were found to be important from hazard and toxicological point of view and those are prevalent in environment are as follows: methylmercury (CH_3Hg^+), ethylmercury ($C_2H_5Hg^+$) [23, 24], phenylmercury ($C_6H_5Hg^+$) [25], and dimethylmercury ($(CH_3)_2Hg$) [26].

2.2. Lead and its organic derivatives

Inorganic lead is introduced into the natural environment from several sources, but organolead compounds are mainly exhausted into the air from the petroleum industry and automobiles, and then they contaminate soil and water sources [27].

Lead organic forms, such as tetramethyl lead and tetraethyl lead have been widely used as antiknocking agents in fuels. Although in the past decade lead gasoline consumption decreased considerably, there are still countries around the globe that use it [28, 29]. Currently, gasoline used in aviation remained the fuel with the highest alkyl lead content, in those days sources of alkyl lead in surrounding environment are airport fuel terminals, bulk aviation, gasoline plants, bulk leaded racing, and other nonroad vehicles gasoline plants, spills from fuel loading, transfer storage, and fuelling [27].

Alkyl lead compounds, such as tetra-alkyl lead are easily absorbed by living organisms due to their lipophilic character. Their absorption depends on the nature of the compound, exposure time, and nature of organism [27, 30]. Toxicological studies on human beings have been demonstrated that cumulative chronic exposure to organic forms of lead is more toxic than those to inorganic forms of lead [31]. According to Gallert and Winter [30] and Pyrzynska [32], the toxicity of alkyl lead compounds decreases with a decreasing number of ethyl or methyl moieties or with a decreasing number of carbon atoms (ethyl lead → methyl lead) according to the following sequence: $R_4Pb > R_3Pb^+ > R_2Pb^{2+}$, with R being either $-CH_3$ or $-C_2H_5$ [27].

2.3. Tin and its organic derivatives

Organotin compounds are organometallic compounds in which carbons are bonded directly to tin (R_nSnX_{4-n}, where n is between 1 and 4, and R is an alkyl or aryl group) [33]. Organometallic forms of tin have been used as active agents in a wide range of applications in industry, such as stabilizers in the polyvinyl chloride industry, plastic additives production, industrial catalysts, antifouling paints, wood preservatives, and in agriculture as biocide products (insecticides, fungicides, and bactericides) [33–35].

Nowadays, use of organotin compounds as anti-foulant has been banned due to their severe toxic effects on the aquatic organisms [36]. Moreover, use of tributyl tin and triphenyl tin compounds in various industrial applications has raised a great concern in the last decades owing to their serious toxic effects on nontarget organisms when leached into environment even at very low concentrations ($ng \cdot L^{-1}$) [37]. Besides the fact that they are considered as endocrine disruptors among organometallic compounds, they also possess teratogenic properties and can cause disruption to the reproductive function in mammals, as well as could act as hepatoxins, immunotoxins, neurotoxins, and obesogens [35, 38].

In the following sections of this chapter, we will present quantitative and qualitative data about the presence of these compounds in different environmental compartments and biota samples.

3. Organomercury, organotin, and organolead detection from complex environmental and biota samples: local case study

Considering the extensive use from past and their improper disposal, as well the lack of evaluation of possible contaminated sites (from past activities) made that even in our days many of such sites to still being used either for agriculture or for pasture. Without a proper evaluation of contaminants, such as mercury, lead, or tin distribution and speciation in soils, and without an assessment of their risks to animals and humans, exposure to such contaminants could be occurred nowadays. To assess such risks for environment, biota, and public health protection purposes, it is imperative to consider their speciation both in soils, water, and in biota plant.

3.1. Environmental and biota sampling for organomercury, organotin, and organolead monitoring

Soil, water, and vegetable samples were collected from Turda region, Cluj district from the northwestern part of Transylvania (46°34′ and 23°47′E) including Turda town, nearest rural regions and industrial zones—banned chemical factory, Romania (see **Figure 1**). Soil, water,

Figure 1. Sampling site map—rectangle corresponds to industrial sites; oval corresponds to inhabited areas; rhomb corresponds to agricultural sites.

and vegetable samples were collected in a period of March, July, and October for 2 years consecutively. Vegetables included for study were selected based on their edible part contact with different environmental compartments: leafy vegetables (lettuce, spinach, and cabbage), "root" and "bulb" vegetables (carrot, parsnip, onion, and garlic), and fruit vegetables (peas, tomato, and eggplant).

Soil and vegetable sample were collected with metallic collectors, returned to the laboratory in polyethylene bags and stored at −20°C. Before analysis, the samples were spread and dried at ambient temperature, and after drying samples were homogenized and shifted through a 2-mm stainless steel sieve.

Surface and well water were collected in polyethylene bottles excepting the cases when organic mercury species were the target analytes, the case when the samples were collected in Teflon containers in order to avoid metallic compound reaction with the bottle surface. Before all sampling campaign, the sampling bottles were subjected to acid cleaning with HNO_3 in order to remove possible metal impurities from the bottle's wall and to prevent further metal adsorption [39]. All water samples were stored in dark at 4°C until analysis and analyzed in less than 7 days from sampling time.

3.2. Organomercury compounds analysis from soil, water, and vegetable samples

As previous work had shown that the organic species of mercury that were found to be important from hazard and toxicological point of view and those are prevalent in the environment are as follows methylmercury (CH_3Hg^+), ethylmercury ($C_2H_5Hg^+$) [23, 24], phenylmercury ($C_6H_5Hg^+$) [25], and dimethylmercury ($(CH_3)_2Hg$) [26].

3.2.1. Water analysis

According to Cai et al. [40] and with minor modifications, extraction and derivatization of organic forms of mercury were conducted using 20 mL of water sample that was placed in 40 mL amber glass vials sealed with screw caps with polytetrafluoroethylene (PTFE)-coated silicon rubber septum. Noted that 2 g of NaCl with 150 μL of 0.4% $NaBEt_2$ was added to the sample and the pH was set to 4.5 using acid acetic, and then the vials were immediately closed tightly. The derivatization step was acquired during 15 min at 70°C.

Afterward, a 50 μm/30 μm divinylbenzene/carboxen/polydimethylsiloxane (DVB/CAR/PDMS) fiber was exposed to the solution headspace for 20 min, maintaining the same temperature (40°C) and assuring a continuous agitation with a rate of 175 rpm. Finally, the fiber was introduced in the chromatographic injector and the target compounds were thermally desorbed at 260°C for 5 min.

3.2.2. Soil and vegetal sample analysis

According to the method presented by Korbas et al. [41], 5 g of homogenized soil and their respective dried vegetable samples were put in an extraction tube with 1 mL of aqueous H_2SO_4 (14 M, saturated with cupric sulfate), 1 mL of 4 M KBr solution, and 20 mL of toluene. The mixture with samples was shacked for 30 min after that subjected to centrifugation for 15 min at 2000 rpm. The procedure was repeated once under the same condition, after that the collected supernatant organic phases were combined and back extracted with 20 mL of L-cysteine solution (1.5% w/v). The organic phase was separated once again after shaking and centrifugation process (2000 rpm for 15 min). From the obtained organic layer, water was removed using anhydrous Na_2SO_4 and from that 1 μL was injected into gas chromatograph inlet.

3.2.3. Gas chromatography-mass spectrometric analysis

Gas chromatography-mass spectrometric analysis was carried out on a GC Focus DSQ II equipment (Thermo Finnigan) using a TR-5MS capillary column with the following characteristics: 30 m × 0.32 mm i.d., with a 0.25-μm film thickness. The mass spectrometer was operated in an electron impact ionization mode at 70 eV ionizing energy. The GC injection port temperature was set at 280°C while the detector source temperature was set at 250°C. Splitless mode was used for injection of 1 μL volume of extracts. Applied temperature program for column oven started from 80°C (3 min) to 150°C·min^{-1} with a rate of 5°C·min^{-1} and maintained at 150°C for 5 min followed by an increase of 10°C·min^{-1} until 280°C and maintained at this final temperature for 5 min. Identification of the target compounds was done through full scan monitoring mode ranging between 50 and 600 m/z.

3.3. Organolead compounds analysis from soil, water, and vegetable samples

Organolead compounds are found in major environmental compartments not only as a consequence of their use in anthropogenic activity, but also via naturally as a consequence of biomethylation processes. As mentioned earlier, the toxicity of these groups of compounds was widely demonstrated, it is known that tetraethyllead (TEL) is much more toxic to animals [16] while ionic allkylead compound was found to be more toxic to plants [42], with both showing higher toxicity than inorganic lead, mainly due to their liposolubility [43]. Generally, it is accepted that the toxicity of organolead compounds increases with the degree of alkylation, respecting the following sequence tetraethyllead > triethyllead > diethyllead > monoethyllead [30].

Target organolead compounds of this study were tetraethyllead (TEL), followed by its transformation products in environment, as triethyllead (TREL), diethyllead (DEL), and monoethyllead (MEL) resulted from dealkylation reactions having as standard their chlorinated forms.

3.3.1. *Water analysis*

The extraction and derivatization of organic forms of lead is similar to the extraction of organotin species. Shortly, 10 mL of water sample was placed in 40 mL amber glass vials sealed with screw caps with PTFE-coated silicon rubber septum. Noted that 2 g of NaCl with 500 μL of 0.4% $NaBEt_2$ was added to sample and the pH was set at 4.5 using acid acetic, after that the vials were immediately tightly closed. The derivatization step was acquired during 100 min at 40°C.

Afterward, 100-μm polydimethylsiloxane (PDMS) fiber was exposed to the solution headspace for 20 min, maintaining the same temperature (40°C) and ensuring a continuous agitation with a rate of 175 rpm. Finally, the fiber was introduced in the chromatographic injector and the target compounds were thermally desorbed at 260°C for 5 min.

3.3.2. *Soil and vegetable sample analysis*

The extraction of organolead compounds from soil and vegetable samples was acquired ultrasound assisted for 20 min using 10 g of samples and 50 mL of n-hexane. The supernatant was collected and the extraction was repeated once again under the same conditions. Collected supernatants were rotary evaporated until 1 mL. The concentrate was mixed with 300 μL of $NaBET_4$ (2 g $NaBET_4$ in 10 mL ethanol), used as a derivatization agent for detection and quantification of the target organic lead compounds. The obtained extract mix was subjected to gas chromatography-mass spectrometric (GC-MS) analysis.

3.3.3. *Gas chromatography-mass spectrometric analysis*

This was carried out on a GC Focus DSQ II equipment (Thermo Finnigan) using a TR-5MS capillary column with the following characteristics: 30 m × 0.25 mm i.d. with a 0.25-μm film thickness. The mass spectrometer was operated in an electron impact ionization mode at 70 eV ionizing energy. The GC injector was set at 260°C while the detector source temperature was set at 280°C. Splitless mode was used for injection of 1 μL volume of extracts. Applied temperature program for column oven was started from 50°C (5 min) to 100°C·min^{-1} with a rate of 7°C·min^{-1} and maintained at 100°C for 2 min followed then by an increase of 15°C·min^{-1} until 280°C and maintained at this final temperature for 10 min also. The identification of target compounds was done through a full-scan monitoring mode between the range of 50 and 600 m/z.

3.4. Organotin compounds analysis from soil, water, and vegetable samples

In this work, a field study was conducted investigating the pathways of organotins in soil-water environment and their uptake potential in vegetables grown on possible contaminated areas.

Monitored organotin compounds were as follows: monobutyltin trichloride (MBT), monophenyltin trichloride (MPT), diphenyltin dichloride (DPT), dibutyltin dichloride (DBT), tributyltin chloride (TBT), and triphenyltin chloride (TPT).

3.4.1. Water samples analysis

Water samples preparation for analysis were done as presented by Kovacs et al. [39] and Hu et al. [44] with minor modifications as follows: about 20 mL of water sample was placed in 40 mL amber glass vials sealed with screw caps with PTFE-coated silicon rubber septum. Noted that 1.5 g of NaCl, 2 mL of acetate buffer (acetate buffer: 82 g·L^{-1} sodium acetate in water adjusted to pH 5 with acetic acid), and 100 μL of $NaBEt_4$ solution (2 g $NaBEt_4$ in 10 mL ethanol) had been added to water sample and the vials were immediately closed and stirred on a magnetic stirrer at 100 rpm and 20°C. The optimal SPME fiber for organotin compounds extraction was found to be 100 μm polydimethylsiloxane (PDMS)-coated fiber. SPME was performed in the headspace. Use of $NaBEt_4$ solution allowed the organotin compounds to be derivatized *in-situ* and simultaneously extracted into the PDMS phase. The fiber was exposed to the headspace for 30 min under continuous stirring, after that the fiber was withdrawn into the needle of the holder and the SPME was placed into the GC injector at 240°C for 10 min in order to allow target compound desorption.

3.4.2. Soil and vegetal sample analysis

The extraction of organotin compounds from soil and vegetable samples was done according to the method presented by Hu et al. [44] with minor modifications as follows: 5 g of soil and vegetables, respectively, were extracted with 50 mL methanol that contains 2 mL of hydrochloric acid (37%) by ultrasonic shaking at 50°C for 30 min. The extraction was repeated twice with 30 mL methanol containing 1.2 mL hydrochloric acid (37%). The supernatants were combined and transferred to a separation funnel containing 100 mL of 30% (w/v) sodium chloride salt and 50 mL of dichloromethane and shacked manually for 10 min. The extraction procedure was repeated in the same manner, after that the collected organic layers were united and subjected for concentration through a rotary evaporator almost until to dryness. The concentrate was mixed with a pH 5.0 buffer solution (acetate buffer: 82 g·L^{-1} sodium acetate in water adjusted to pH 5 with acetic acid) and 120 °L of $NaBET_4$ (2 g $NaBET_4$ in 10 mL ethanol) was used as a derivatization agent for the target organic tin compounds detection and quantification. The ethylated organotin compounds were extracted with 5 mL of hexane three times after that the extracts were cleaned up using florisil column. The collected organic fraction was concentrated at 1 mL after that the obtained extracts were subjected to gas chromatography-mass spectrometric (GC-MS) analysis.

3.4.3. Gas chromatography-mass spectrometric analysis

It was carried out on a GC Focus DSQ II equipment (Thermo Finnigan) using a TR-5MS capillary column with the following characteristics: 60 m × 0.25 mm i.d. with a 0.25-μm film thickness. The mass spectrometer was operated in an electron impact ionization mode at 70 eV ionizing energy. The GC injector was set at 270°C while the detector source temperature was set at 280°C. Splitless mode was used for injection of 1 μL volume of extracts. Applied temperature program for column oven was started from 50°C (5 min) to 130°C·min^{-1} with a rate of 15°C·min^{-1} and maintained at 130°C for 10 min followed then by an increase of 15°C·min^{-1} until 280°C and maintained at this final temperature for 10 min.

Identification of target compounds was done through a selective ion-monitoring mode, and the fragment ions were those most abundant in each oligomers. Their values are presented in **Table 1**.

Ions monitored in SIM mode for organotin compounds

Organotin compounds	Primary characteristic ion* (m/z)	Secondary ions (m/z)
MBT	149	179, 233, 235
DBT	149	179, 207, 263
TBT	149	177, 207, 263
MPT	195	253, 255
DPT	275	301, 303
TPT	197	349, 351

*Quantitation ion.

Table 1. Monitored ions in selective ion monitoring mode (SIM) of target organometallic compounds.

4. Organomercury, organotin, and organolead and their "fingerprint" on environmental compartments and surrounding biota: case study, Turda region

The presence of these organometallic compounds surrounding us has raised a great deal of concern in the past few decades because of their high toxicity to nontarget organisms when leached into different environmental compartments, even at low concentrations [2, 23, 33]. At these moments, most studies regarding organometallic compounds present in the environment were conducted on aquatic environment. Studies considering soil or air contamination with organometallic compounds were most of the time conducted on industrial sites and mining areas. Similarly, although studies on bioconcentration and biomagnification of organometallic compounds were done on a large scale of species from the aquatic environment, according to our knowledge, there is minor information in the literature considering organometallic compounds uptake potential by vegetables.

In this work, a field study was conducted investigating the pathways of organometallic compounds (tin, lead, and mercury species) in soil-water environment and their uptake potential in vegetables grown on inhabited area—Turda region, including rural, urban, and agricultural sites considering these sites as possible contaminated areas due to the presence of an old chemical production plant (chemicals factory Turda) which was closed from more than 20 years.

4.1. Organic compounds analysis in environmental samples: case study, Turda region—Cluj, Romania

Surface water as well as underground water samples were collected from the mentioned sites (rural, urban, agricultural, and industrial areas) in order to evaluate if there are any signs considering the potential of past pollution. Considering surface water samples, total 22 sampling points were selected for surface water status evaluation. Specifically, 7 sampling points were selected located on the banned industrial sites, 10 sampling points from inhabited areas (3 sampling points from Turda town and 7 sampling points from rural areas), and 5 sampling points from the agricultural place. Wells water, totally from 21 sampling sites, were collected as follows: 3 sampling points were selected located on the banned industrial sites, 14 sampling points from inhabited areas (4 sampling points from Turda town and 10 sampling points from rural areas), and 4 sampling points from agricultural place.

Organometallic compounds detected in water samples including surface and well water (underground water) showed higher levels in the samples collected from old industrial site than in the other sampling locations (inhabited area—rural and urban, and agricultural sites). The range of organotin, organolead, and organomercury amount detected in the water samples vary between 0.1 and 72, 0.2 and 15.9, and 0.08 and 61.4 $\mu g \cdot L^{-1}$, respectively.

Considering **Figure 2**, it could be observed that the organometallic compounds that were detected in the higher amount in the samples collected from the banned industrial area were methylmercury, ethylmercury, tetraethyllead, and triphenyltin with the average amounts of 26.8, 11.4, 19.5, and 37.7 $\mu g \cdot L^{-1}$, respectively.

Excluding amount value of water samples collected from the banned industrial area, higher values, with 15–20%, of monitored organometallic compounds were observed in the case of underground water samples than those of surface water samples.

Compounds, such as dimethylmercury, monoethyllead, monobutyltin, monophenyltin, and diphenyltin were detected just in 41% in the collected water samples. This could be attributed to their instability considering their physicochemical properties. Taking into account the variations between the period of sampling (March, July, and October), no significant differences were observed.

http://dx.doi.org/10.5772/67755

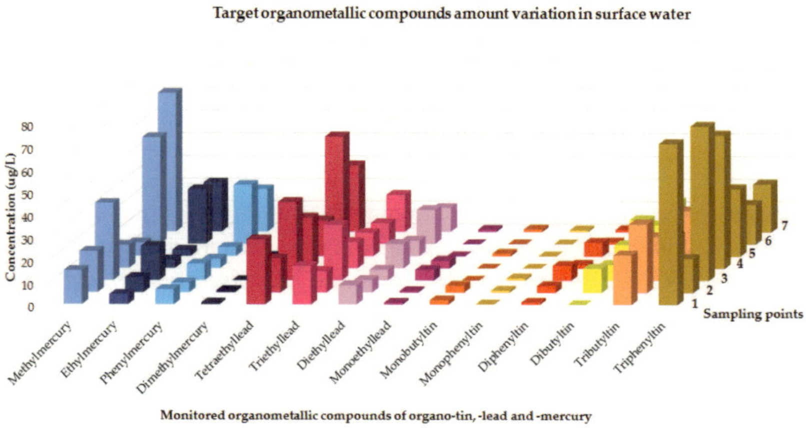

Figure 2. Target organometallic compounds amount ($\mu g \cdot L^{-1}$) in surface water sampling points from industrial sites .

Analysis of soil samples, totally 29 (inhabited areas: 7 sampling points from Turda town and 14 sampling points from villages; 5 from agricultural areas; and 3 from banned industrial sites) has shown higher amounts than analysis from water samples. Soil samples were collected from horizon 0, and from 15 and 30 cm depth, respectively. Between the layers, minor decreasing tendency was observed once with decreasing the depth.

Comparing detected amounts of organomercury compounds of soil samples from a rural area with those of wells water from the same region (**Figure 3**), it was observed that the average value of all four monitored organomercury compounds were approximately twofold higher in the case of soil samples than in well water (underground water).

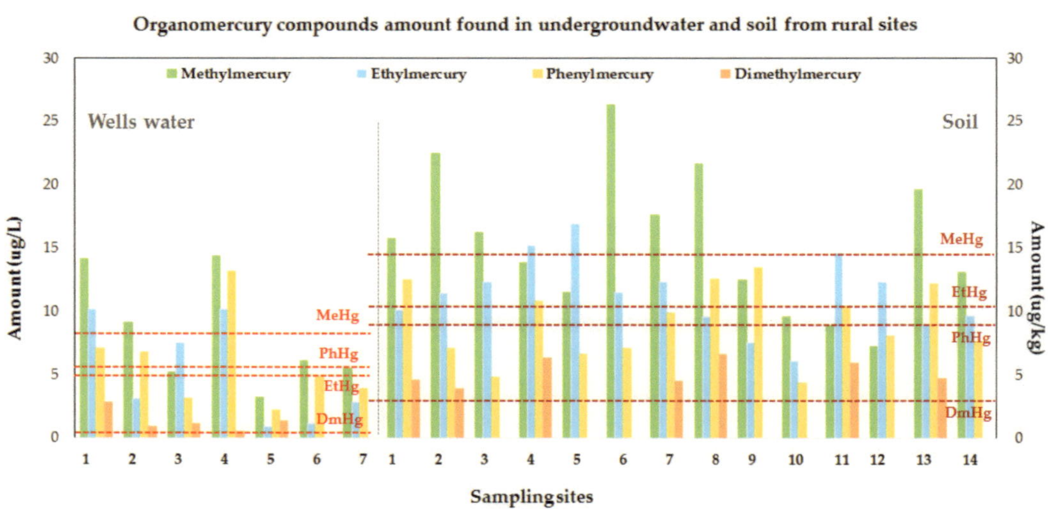

Figure 3. Organomercury compound variation in underground water (wells) and soil from rural sites.

Organotin compounds were detected at a higher percentage in samples from the industrial area compared to the rest of sampling locations (urban, rural, or agricultural areas). Monobutyltin and monophenyl tin were detected just in 40% of the collected samples—including water and soil samples from all sampling sites. This could be caused by their weak stability in the environment.

According to **Figure 4**, the average values of organotin compounds increase once with increasing their degree of alkylation in both the case of water as well as soil samples. In soil samples collected on depth layers, in most cases the organotin compounds were extremely lower in the surface horizons and the amount starts to increase once with depth (see **Figure 4**).

The total amount of organotin compounds in water and soil samples collected from the banned industrial site vary between 38 and 364.1 ppm while from the rest of sampling location the total amount of organotin compounds vary between 3.4 and 21.5, 1.6 and 31.5, and 1.4 and 28.9 ppm.

With regard to organolead compounds, no correlation was found between the detected amounts and sampling locations, as shown in **Table 2**.

4.2. Organometallic compounds uptake by vegetables grown on possible contaminated sites: case study, Turda region—Cluj, Romania

The major pathway of living organism's exposure to contaminants in soil is through food ingestion (food web). The prediction or estimation of risks to living organisms requires knowledge about both environmental contamination and food contamination. The direct measurement of contaminant concentration in living organisms' food is advisable to minimize uncertainty in ecological risk assessment [45]. However, site-specific data on the bioaccumulation of contaminants in vegetation and other biota that comprise living organisms diet are often not available because of constraints in funding and/or time, most of the analysis being expensive and time-consuming.

Most time, the concentration of chemical contaminants is measured at the known contaminated sites prior to a risk assessment. The challenge for the development of numerical models that are able to estimate the concentration of contaminants accumulated in plants (as vegetables, crops, etc.) is the soil-plant uptake factor. Soil-plant uptake factor refers to the ratio of a contaminant in plants to that in soil. The concentration of a contaminant in plant at a particular location is estimated by multiplying the measured concentration in soil by the soil-plant uptake factor [45].

Usually, the concentrations of monitored contaminant in mature plants and soil are assumed to be at equilibrium, thus exposure time is not necessary to be considered. However, in nature the bioavailability of organometallic compounds depend on the geochemical nature, pedoclimatic variables (temperature and rain intensity) and related to fluctuations of physicochemistry of the medium, such as soil moisture, pH, and soil organic matter content [46]. On the other side, the bio-uptake process depends on the internalization pathways of plant species which refer to the ability of organometallic compounds to cross the biological barrier. That, most of the time, is determined by the concentration of flux of internalized organometallic compounds but studied has showed that this ability of plants depends also on their size, nature, and physiology.

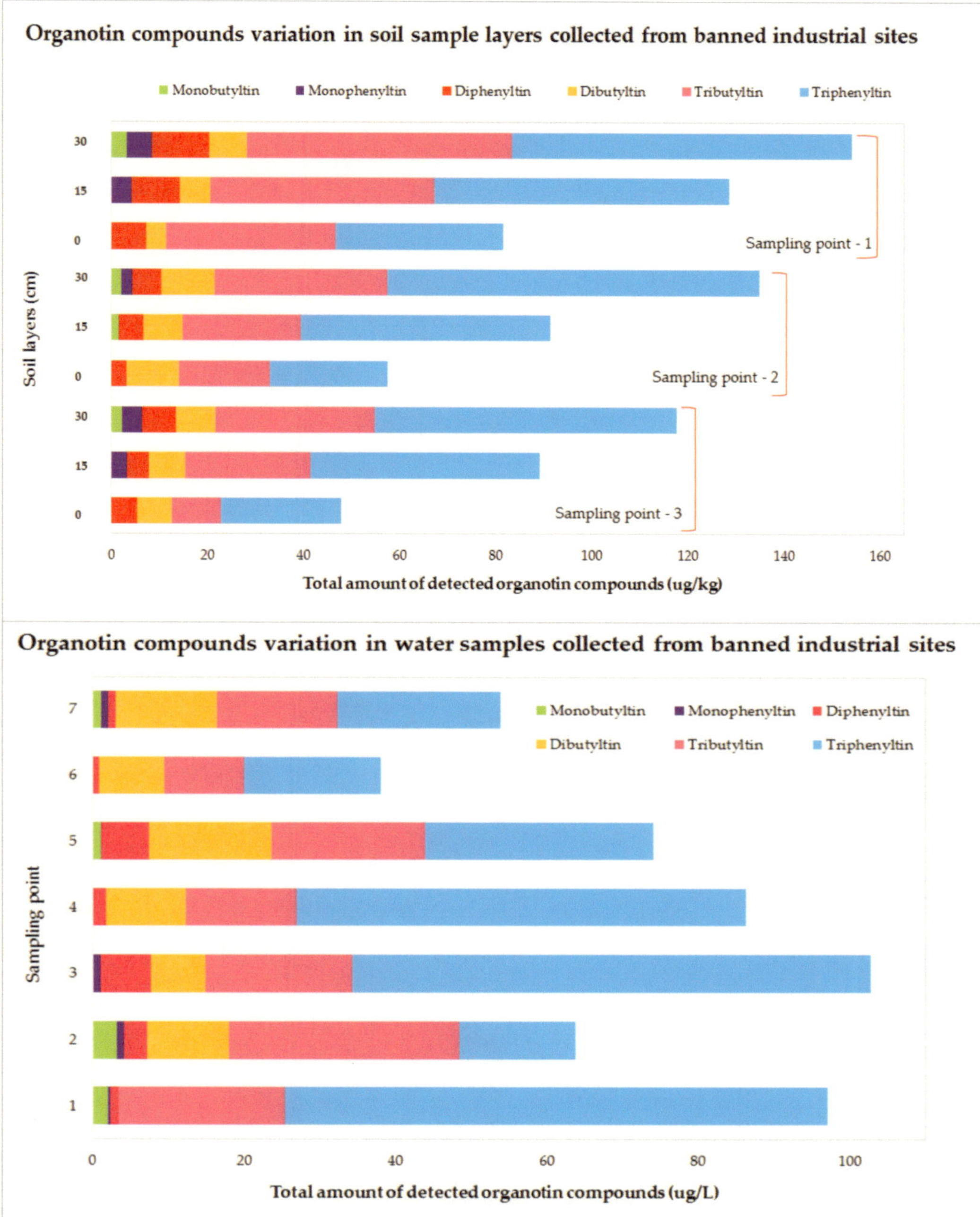

Figure 4. Organotin compound variation between soil (layers) and water samples collected from the sampling points of banned industrial sites.

The uptake factor of monitored organometallic compounds varied between plants (considering their edible part contact with contaminated environmental matrix), see **Table 3**. With respect to their fates in relationship to plant uptake, organometallic compounds in soil can be roughly divided into three groups according to their uptake factors: lowest average uptake

Organolead compounds			TEL			TREL			DEL			MEL		
	Sample type	No.	Range	Average	N.S.U.D.	Range	Average	N.S.U.D.	Range	Average	N.S.U.D.	Range	Average	N.S.U.D.
Banned industrial site	Soil (μg·kg⁻¹)	3	0.1–34.2	12.8	–	0.5–21.5	8.6	–	0.4–7.2	3.8	1	0.5–4.9	0.8	1
	Surface water (μg·L⁻¹)	7	0.2–18.5	5.9	3	0.3–15.5	6.7	2	0.4–10.8	3.9	2	0.4–5.1	1.2	4
	Well waters (μg·L⁻¹)	3	0.2–25.8	11.5	1	0.4–18.2	7.1	-	0.3–6.9	2.4	1	0.5–3.8	0.9	2
Urban site	Soil (μg·kg⁻¹)	7	0.5–10.2	3.5	2	1.2–13.5	6.5	1	0.5–11.8	3.2	2	1.5–5.9	1.8	2
	Surface water (μg·L⁻¹)	3	0.2–8.7	2.9	–	0.8–9.2	2.8	–	0.2–5.9	1.8	1	0.4–3.5	0.7	1
	Well waters (μg·L⁻¹)	4	0.4–12.6	4.5	–	0.4–5.5	3.1	–	0.4–9.7	2.3	1	0.2–4.9	2.5	1
Rural site	Soil (μg·kg⁻¹)	14	0.3–21.8	10.5	3	0.5–16.2	5.8	2	0.2–7.4	3.8	4	1.2–6.2	2.6	6
	Surface water (μg·L⁻¹)	7	1.2–15.5	5.5	2	0.2–6.7	2.1	1	0.4–4.2	1.5	2	0.4–3.9	0.8	3
	Well waters (μg·L⁻¹)	10	2.8–16.7	8.2	4	0.4–15.5	3.9	3	0.4–6.1	3.1	2	0. –6.2	3.5	4

Organolead compounds			TEL			TREL			DEL			MEL		
	Sample type	No.	Range	Average	N.S.U.D.	Range	Average	N.S.U.D.	Range	Average	N.S.U.D.	Range	Average	N.S.U.D.
Agricultural site	Soil ($\mu g{\cdot}kg^{-1}$)	5	1.5–20.4	7.7	–	0.1–13.2	7.5	1	0.2–13.5	5.2	2	0.5–4.8	1.6	2
	Surface water ($\mu g{\cdot}L^{-1}$)	5	0.5–13.2	3.5	–	0.2–6.8	1.8	1	0.4–7.9	3.8	1	0.3–2.8	1.1	2
	Well waters ($\mu g{\cdot}L^{-1}$)	4	0.2–13.5	8.5	–	0.2–9.5	4.2	–	1.2–5.8	2.1	–	0.2–3.7	0.9	1

Notes: TEL, tetraethyllead; TREL, triethyllead; DEL, diethyllead; MEL, monoethyllead; No, –number of sample collected; N.S.U.D, number of samples in which the target compounds was not detected.

Table 2. Organolead species variation between sampling sites.

Organometallic compounds	Leafy vegetables (n = 102)		Root and bulb vegetables (n = 115)		Fruit vegetables (n = 95)	
	Average	Min/Max.	Average	Min/Max.	Average	Min/Max.
Methylmercury	0.64	0.02–1.05	0.84	0.07–1.57	0.57	0.02–1.81
Ethylmercury	0.74	0.04–1.84	0.94	0.04–1.84	0.61	0.004–1.37
Phenylmercury	0.48	0.007–1.15	1.14	0.008–1.61	0.24	0.005–0.98
Dimethylmercury	0.44	0.08–1.24	1.06	0.004–1.28	0.15	0.02–0.87
Tetraethyllead	1.24	0.004–1.51	1.15	0.08–1.47	0.44	0.002–0.98
Triethyllead	1.05	0.002–1.61	0.95	0.003–1.27	0.51	0.01–1.05
Diethyllead	0.42	0.05–0.98	0.61	0.005–1.36	0.38	0.002–0.74
Monoethyllead	0.34	0.04–0.54	0.27	0.05–0.84	0.18	0.002–0.44
Monobutyltin	0.28	0.006–0.49	0.19	0.001–0.31	0.17	0.001–0.39
Monophenyltin	0.18	0.002–0.37	0.11	0.002–0.41	0.24	0.002–0.58
Diphenyltin	0.46	0.001–0.49	0.24	0.005–0.64	0.33	0.01–0.77
Dibutyltin	0.22	0.005–0.57	0.41	0.001–0.81	0.19	0.002– 0.69
Tributyltin	1.24	0.002–1.65	0.67	0.004–0.79	0.47	0.02–1.09
Triphenyltin	1.09	0.05–1.74	0.53	0.002–1.15	0.61	0.008–1.11

n = number of vegetable samples.

Table 3. Uptake factors range for monitored organometallic compounds.

http://dx.doi.org/10.5772/67755

factor (between 0.001 and 0.05), medium average uptake factor (0.05–0.1), and highest average uptake factor (0.1–1). The highest uptake factors for almost all monitored organometallic compounds were determined in the case of root and bulb vegetables, flowed by leafy vegetables, and finally by fruit vegetables (see **Table 3**).

Lowest uptake factors were determined in the case of monoethyllead, monobutyltin, monophenyltin, diphenyltin, and dibutyltin regardless of plant species group (leafy, root, bulb, fruit, and vegetables).

5. Summary

The problem of trace metals pollution is transformation of the original pollutants into other species as organometallic compounds of mercury, tin, and lead species that are considered frequently more toxic and mobile in the environment than their corresponding parent compound. Although organometallic compounds occurrence in the environment could be natural, most of the time their quantitative presence is associated with anthropogenic sources and activities. Organometallic amounts in detected environmental samples increased once with their degree of alkylation in the case of all three groups studied (tin, lead, and mercury).

Based on the literature data, worldwide observations showed that a clear relation is observed between organometallic compound concentrations in the soils/underground water and the accumulation of this element in plants. In the case of monitored organometallic compounds, the higher amounts were detected in root and bulb vegetables followed by leafy, vegetables, and fruit vegetables. The uptake factor increased once with the degree of alkylation in the case of all groups studied.

Author details

Kovacs Melinda Haydee[1*] and Kovacs Emoke Dalma[1,2]

*Address all correspondence to: haydee.kovacs.melinda@gmail.com

1 Research Institute for Analytical Instrumentation, INCDO-INOE 2000, Cluj-Napoca, Romania

2 Chemistry and Chemical Engineering Faculty, Babes-Bolyai University, Cluj-Napoca, Romania

References

[1] Ding Q, Cheng G, Wang Y, Zhuang D. Effects of natural factors on the spatial distribution of heavy metals in soils surrounding mining regions. Science of the Total Environment. 2017;**578**:577-585

[2] Lee R, Oshima Y. Effects of selected pesticides, metals and organometallics on development of blue crab (*Callinectes sapidus*) embryos. Marine Environmental Research. 1998;**46**(1-5):479-482

[3] Office of the Science Advisor – Risk Assessment Forum, editor. Framework for Metal Risk Assessment. Washington DC, USA: U.S. Environmental Protection Agency; 2007. p. 172. DOI: EPA 120/R-07/001

[4] Wu SH, Shi YX, Zhou SL, Wang CH, Chen H. Modelling and mapping of critical loads for heavy metals in Kunshan soil. Science of the Total Environment. 2016;**569-570**:191-200

[5] Henriques B, Rodrigues SM, Coelho C, Cruz N, Duarte AC, Romkens PFAM, Pereira E. Risks associated with the transfer of toxic organo-metallic mercury from soils into the terrestrial feed chain. Environmental International. 2013;**59**:408-417

[6] Hu H, Li Z, Feng Y, Liu Y, Xue J, Davis M, Liang Y. Prediction model for mercury transfer from soil to corn grain and its cross-species extrapolation. Journal of Integrative Agriculture. 2016;**15**(10):2393-2402

[7] Pan LB, Ma J, Wang XL, Hou H. Heavy metals in soils from a typical county in Shanxi Province, China: Levels, sources and spatial distribution. Chemosphere. 2016;**148**:248-254

[8] Aran D, Gury M, Jeanroy E. Organo-metallic complexes in an Andosol: A comparative study with a Cambisol and Podzol. Geoderma. 2001;**97**(1-2):65-79

[9] Lu M, Wu XJ, Zheng DC, Liao Y. Distribution of PCDD/Fs and organometallic compounds in sewage sludge of wastewater treatment plants in China. Environmental Pollution. 2012;**171**:78-84

[10] Zhaou YJ, He JC, Chen Q, He J, Hou HQ, Zheng Z. Evaluation of 206 nm UV radiation transformation for degrading organometallics in wastewater. Chemical Engineering Journal. 2011;**167**(1):22-27

[11] Reis AT, Rodriques SM, Davidson CM, Pereira E, Duarte AC. Extractability and mobility of mercury from agricultural soils surrounding industrial and mining contaminated areas. Chemosphere. 2010;**81**(11):1369-1377

[12] Biester H, Muller G, Scholer HF. Binding and mobility of mercury in soils contaminated by emissions from chloralkili plants. Science of the Total Environment. 2002;**284**:191-203

[13] Selin N.E. Global biogeochemical cycling of mercury: A review. Annual Review of Environment and Resources. 2009;**34**:43-63

[14] Challa YR, Astudillo LRD, Ramirez A, Escalona A, Martinez G. Distribution of total and organic mercury in superficial soils in the upon Manzanares River watershed, Sucre State, Venezuela. Air, Soil and Water Research. 2008;**1**:21-29

[15] Clarckson TW, Magos L. The toxicology of mercury and its chemical components. Critical Review in Toxicology. 2006;**36**:609-662

[16] Craig PJ, Eng G, Jenkins RO. Occurrence and pathways of organometallic compounds in the environment – general considerations. In: Craig PJ, editor. Organometallic Compounds Occurrence in the Environment. 2nd ed. England: John Wiley & Sons Ltd; 2003. pp. 1-56

[17] Briant N, Chouvelon T, Martinez L, Brach-Papa C, Chiffoleau JF, Savoye N, Sonke J, Knoery J. Spatial and temporal distribution of mercury and methylmercury in bivalves from the French coastline. Marine Pollution Bulletin. 2017;**114**(2):1096-1102

[18] Chen L, Liu M, Fan R, Ma S, Xu Z, Ren M, He Q. Mercury speciation and emission from municipal solid waste incinerators in the Pearl River Delta, South China. Science of the Total Environment. 2013;**447**:396-402

[19] Caballero B, Olguin N, Campos F, Farina M, Ballester F, Lopez-Espinosa MJ, et al. Methylmercury-induced developmental toxicity is associated with oxidative stress and cofilin phosphorylation. Cellular and human studies. NeuroToxicology. 2017;**59**:197-209. DOI: http://dx.doi.org/10.1016/j.neuro.2016.05.018

[20] Ser PH, Omi S, Shimizu-Furusawa H, Yasutake A, Sakamoto M, Hachiya N, et al. Differences in the responses of three plasma selenium-containing proteins in relation to methylmercury-exposure through consumption of fish/whales. Toxicology Letters. 2017;**267**:53-58

[21] Morita M, Yoshinaga J, Edmonds JS. The determination of mercury species in environment and biological samples (Technical Report). In: Analytical Chemistry Division - Commission of Microchemical Techniques and Trace Analysis Working Group on Speciation, editors. Pure and Applied Chemistry. 1st ed. Great Britain: IUPAC; 1998. pp. 1585-1615

[22] Lindqvist O, Jernelov A, Johansson K, Rodhe H. Mercury in the Swedish environment: Global and local sources. In: Report of the Workshop; November 1983; Lerum, Sweden. Solna, Sweden: SNV PM I Swedish Environmental Protection Board; 1984. pp. 105-110

[23] Marques RC, Abreu L, Bernardi JVE, Dorea JG. Neurodevelopment of Amazonian children exposed to ethylmercury (from Thimerosal in vaccines) and methylmercury (from fish). Environmental Research. 2016;**149**:259-265

[24] Zimmermann LT, Santos DB, Naime AA, Leal RB, Dorea JG, Barbosa Jr F, et al. Comparative study on methyl- and ethylmercury-induced toxicity in C6 glioma cells and the potential role of LAT-1 in mediating mercurial-thiol complexes uptake. NeuroToxicology. 2013;**38**:1-8

[25] Lee CH, Lin RH, Liu SH, Lin-Shiau SY. Distinct genotoxicity of phenylmercury acetate in human lymphocytes as compared with other mercury compounds. Mutation Research/Genetic Toxicology and Environmental Mutagenesis. 1997;**392**(3):269-276

[26] Siegler WR, Nierenberg DW, Hickey WF. Fatal poisoning from liquid dimethylmercury: A neuropathologic study. Human Pathology. 1999;**30**(6):720-723

[27] Zuniga MC, Jover E, Arancibia V, Bayona JM. Development of a methodology for the simultaneous determination of inorganic and organolead compounds using supercritical fluid extraction followed by gas chromatography-mass spectrometry and its application to environmental matrices. Talanta. 2009;**80**(2):504-510

[28] Elizabeth O'Brien/The Lead Group Inc. The Lead Education and Abatement Design Group Working to Eliminate Lead Poisoning Globally and to Protect the Environment from Lead in all its Uses: Past, Current and New Uses ABN 25 819 463 114 [Internet]. 2007. Available from: http://www.lead.org.au/fs/fst27_20070524.html [Accessed: January 10,2017]

[29] Asian Clean Fuels Association. Africa fuel quality update – progress & challenges. Asian Clean Fuels Association (ACFA News). 2010;**8**(5):1-6

[30] Gallert C, Winter J. Bioremediation of soil contaminated with alkyllead compounds. Water Research. 2002;**36**(12):3130-3140

[31] Blakey DH, Bayley JM, Douglas GR. Induction of chromosomal aberrations in Chinese hamster ovary cells by triethyllead acetate. Mutation Research/Genetic Toxicology. 1992;**298**(1):1-7

[32] Pyrzynska K. Organolead speciation in environmental samples: A review. Microchimica Acta. 1996;**122**(3):279-293

[33] Gao JM, Wu L, Chen YP, Zhou B, Guo JS, Zhang K, Ouyang WJ. Spatiotemporal distribution and risk assessment of organotins in the surface water of the Three Gorges Reservoir Region, China. Chemosphere. 2017;**171**:405-414

[34] Hoch M. Organotin compounds in the environment – An overview. Applied Geochemistry. 2001;**16**:719-743

[35] Cole RF, Mills GA, Parker R, Bolam T, Birchenough A, Kroger S, Fones GR. Trends in the analysis and monitoring of organotins in the aquatic environment. Trends in Environmental Analytical Chemistry. 2015;**8**:1-11

[36] Birchenough AC, Barnes N, Evans SM, Hinz H, Kronke I, Moss C. A review and assessment of tributyltin contamination in the North Sea, based on surveys of tributyltin tissue burdens and imposex/intersex in the four species of neogastropods. Marine Pollution Bulletin. 2002;**44**:534-543

[37] Nakanishi. Endocrine disruption induced by organotin compounds; organotins function as a powerful agonist for nuclear receptors rather than an aromatase inhibitor. The Journal of Toxicological Sciences. 2008;**33**:269-276

[38] Zuo Z, Chen S, Wu T, Zhang J, Su Y, Chen Y, Wang C. Tributyltin causes obesity and hepatic steatosis in male mice. Environmental Toxicology. 2011;**26**(1):79-85

[39] Kovacs MH, Ristoiu D, Voica C, Ristoiu T. Optimization of organometallic compounds extraction from aqueous samples in order to improve their gas chromatography-mass spectrometry analysis performance. Romanian Journal of Physics. 2013;**58**(1-2):204-210

[40] Cai Y, Monsalud S, Jaffe R, Jones RD. Gas chromatographic determination of organomercury following aqueous derivatization with sodium tetraethylborate and sodium tetraphenylborate: Comparative study of gas chromatography coupled with atomic fluorescence spectrometry, atomic emission spectrometry and mass spectrometry. Journal of Chromatography A. 2000;**876**(1-2):23-34

[41] Korbas M, Blechinger SR, Krone PH, Pickering IJ, George GN. Localizing organomercury uptake and accumulation in zebrafish larvae at the tissue and cellular level. Proceedings of the National Academy of Science of the United States of America. 2008;**105**(34):12108-12112

[42] Fallon RD. Accelerated rates of organolead transformation following nutrient enrichment of contaminated ground water. Bulletin of Environmental Contamination and Toxicology. 1994;**53**(4):603-609

[43] Penalver R, Campillo N, Hernandez-Cordoba M. Comparison of two derivatization reagents for the simultaneous determination of organolead and organomanganese compounds using solid-phase microextraction followed by gas chromatography with atomic emission detection. Talanta. 2011;**87**:268-275

[44] Hu JJ, Zhen H, Wan Y, Gao J, An W, An L, Jin F, Jin X. Trophic magnification of triphenyltin in a marine food web of Bohai Bay, North China: Comparison to tributyltin. Environmental Science and Technology. 2006;**40**(10):3142-3147

[45] Office of Environmental Management, Bechtel Jacobs Company LLC, editors. Empirical models for the uptake of inorganic chemicals from soil by plants. 1st ed. Oak Ridge National Laboratory, USA: East Tennessee Technology Park; 1998. p. 116

[46] Temmerman LD, Waegeneers N, Thiry C, Laing GD, Tack F, Ruttens A. Selenium content of Belgian cultivated soils and its uptake by field crops and vegetables. Science of the Total Environment. 2014;**468-469**:77-82

Section 2

State-of-the-Art: Catalysis

Chapter 4

Triple Bonds between Bismuth and Group 13 Elements: Theoretical Designs and Characterization

Jia-Syun Lu, Ming-Chung Yang, Shih-Hao Su,
Xiang-Ting Wen, Jia-Zhen Xie and Ming-Der Su

Additional information is available at the end of the chapter

http://dx.doi.org/10.5772/67220

Abstract

The effect of substitution on the potential energy surfaces of $RE_{13}{\equiv}BiR$ (E_{13} = B, Al, Ga, In, and Tl; R = F, OH, H, CH_3, SiH_3, Tbt, Ar*, SiMe(Si*t*Bu_3)$_2$, and Si*i*PrDis$_2$) is investigated using density functional theories (M06-2X/Def2-TZVP, B3PW91/Def2-TZVP, and B3LYP/LANL2DZ+dp). The theoretical results suggest that all of the triply bonded $RE_{13}{\equiv}BiR$ molecules prefer to adopt a bent geometry (i.e., $\angle RE_{13}Bi \approx 180°$ and $\angle E_{13}BiR \approx 90°$), which agrees well with the bonding model (model (B)). It is also demonstrated that the smaller groups, such as R = F, OH, H, CH_3, and SiH_3, neither kinetically nor thermodynamically stabilize the triply bonded $RE_{13}{\equiv}BiR$ compounds, except for the case of $H_3SiB{\equiv}BiSiH_3$. Nevertheless, the triply bonded $R'E_{13}{\equiv}BiR'$ molecules that feature bulkier substituents (R' = Tbt, Ar*, SiMe(Si*t*Bu_3)$_2$, and Si*i*PrDis$_2$) are found to have the global minimum on the singlet potential energy surface and are both kinetically and thermodynamically stable. In other words, both the electronic and the steric effects of bulkier substituent groups play an important role in making triply bonded $RE_{13}{\equiv}BiR$ (Group 13–Group 15) species synthetically accessible and isolable in a stable form.

Keywords: bismuth, group 13 elements, triple bond, multiple bond, density functional theory

1. Introduction

Triply bonded molecules are of great interest in structural and synthetic inorganic chemistry as well as in fundamental science. Molecules that have triple bonds, however, pose a more difficult challenge than analogous doubly bonded molecules from a synthetic viewpoint [1–8]. Acetylene is one of the most commonly triply bonded molecules in traditional organic chemistry. Thanks to Kira, Power, Sekiguchi, Tokitoh, Wiberg and many coworkers, the stable homonuclear alkyne analogues of all of the heavier group 14 elements have now been isolated and characterized [9–19]. Recently, heteronuclear ethyne-like molecules that possess C≡Ge [20, 21], C≡Sn [22], and C≡Pb [23] triple bonds have also been theoretically predicted and have been published elsewhere.

Nevertheless, to the authors' best knowledge, neither experimental nor theoretical studies have been performed on acetylene-like compounds that feature an E_{13}≡Bi (E = B, Al, Ga, In, and Tl) triple bond. It is surprising how little is known about the stability and molecular properties of E_{13}≡Bi, considering the importance of bismuth compounds [24] that contain group 13 elements in inorganic chemistry [25–35] and material chemistry [36–45].

The aim of this study is to theoretically determine the existence and relative stability of RE_{13}≡BiR triply bonded molecules, which can be synthesized as stable compounds when they are properly substituted. For the first time, the structures of RE_{13}≡BiR with various substituents are reported. That is, theoretical calculations of RE_{13}≡BiR are performed, using both smaller ligands (such as, R = F, OH, H, CH_3, and SiH_3) and larger ligands with bulky aryl and silyl groups (i.e., R' = Tbt, Ar*, SiMe(Si*t*Bu$_3$)$_2$, and Si*i*PrDis$_2$; Dis = CH(SiMe$_3$)$_2$; **Scheme 1**) [46–51]. As a result, the effect of substituents on these bismuth-group-13-element triple bonds is systematically investigated using density functional theory (DFT) calculations. It is expected that the theoretical interpretations of the effect of substituents, presented in this work, will help in the experimental preparation of the many precursors of RE_{13}≡BiR.

Tbt **Ar***

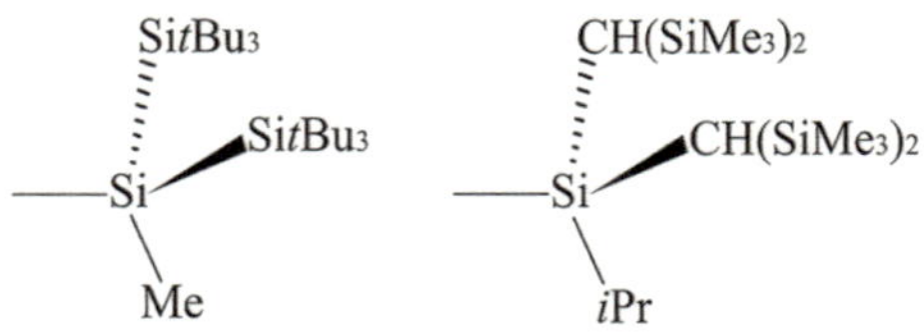

SiMe(Si*t*Bu$_3$)$_2$ **Si*i*PrDis$_2$**

2. Theoretical methods

Geometries were fully optimized using hybrid density functional theory at M06-2X, B3PW91, and B3LYP levels, using the Gaussian 09 program package [52]. It has been reported that M06-2X is proven to have excellent performance for main group chemistry [53]. In both the B3LYP and B3PW91 calculations, Becke's three-parameter nonlocal exchange functional (B3) [54, 55] is used, together with the exact (Hartree-Fock) exchange functional, in conjunction with the nonlocal correlation functional of Lee, Yang, Parr (LYP) [56] and Perdew and Wang (PW91) [57]. Therefore, the geometries of all of the stationary points were fully optimized at the M06-2X, B3PW91, and B3LYP levels of theory. For comparison, the geometries and energetics of the stationary points on the potential energy surface were calculated using the M06-2X, B3PW91, and B3LYP methods, in conjunction with the Def2-TZVP [58] and LANL2DZ+dp [59–62] basis sets. Consequently, these DFT calculations are denoted as M06-2X/Def2-TZVP, B3PW91/Def2-TZVP, and B3LYP/LANL2DZ+dp, respectively.

The spin-unrestricted (UM06-2X, UB3PW91, and UB3LYP) formalisms are used for the open-shell (triplet) species. The $\langle S^2 \rangle$ expectation values for the triplet state for the calculated species all have an ideal value (2.00), after spin annihilation, so their geometries and energetics are reliable for this study. Frequency calculations were performed on all structures, in order to confirm that the reactants and products have no imaginary frequencies, and that the transition states possess only one imaginary frequency. Thermodynamic corrections to 298 K, heat capacity corrections, and entropy corrections (ΔS) are applied at the three DFT levels. Therefore, the relative free energy (ΔG) at 298 K is also calculated at the same levels of theory.

Sequential conformation analyses were performed for each stationary point, for species containing bulky ligands (R' = Tbt, Ar*, SiMe(Si*t*Bu$_3$)$_2$, and Si*i*PrDis$_2$) using Hartree-Fock calculations (RHF/3-21G*). The TbtE$_{13}$≡Bi=Tbt, Ar*=E$_{13}$≡Bi=Ar*, SiMe(Si*t*Bu$_3$)$_2$=E$_{13}$≡Bi=SiMe(Si*t*Bu$_3$)$_2$, and Si*i*PrDis$_2$=E$_{13}$≡Bi=Si*i*PrDis$_2$ (E =B, Al, Ga, In, and Tl) are used as model reactants in this work. It is known that the Hartree-Fock level of theory is insufficient for even a qualitative description of the chemical potential energy surface, so these stationary points were then further calculated at the B3LYP/LANL2DZ+dp level, using the OPT=READFC keyword with a tight convergence option (maximum gradient convergence tolerance = 5.0×10^{-5} hartree/bohr). Because of the limitations of the available CPU time and memory size, frequencies were not calculated for the triply bonded R'E$_{13}$≡BiR' systems with bulky ligands (R') at the B3LYP/LANL2DZ+dp level of theory. As a result, the zero-point energies and the Gibbs free energies for B3LYP/LANL2DZ+dp cannot be applied to these systems.

3. Results and discussion

3.1. Theoretical models for RE$_{13}$≡BiR

In order to understand the bonding interactions in the R=E$_{13}$≡Bi=R molecule, R=E$_{13}$≡Bi=R is divided into one E$_{13}$=R and one Bi=R fragment. The theoretical calculations for these two fragments indicate that the ground states of E_{13}=R and Bi=R are singlet and triplet states, respectively (vide infra). Therefore, there are two possible interaction modes (A and B) between the E_{13}=R and Bi=R moieties in the formation of the triply bonded R=E_{13}≡Bi=R species, as schematically illustrated in **Figure 1**. In model (A), both E=R and Bi=R units exist as triplet monomers. In this way, the combination between the group 13 element and bismuth can be considered as a triple bond, since it consists of 2 π bonds and 1 donor-acceptor σ bond, for these 2 triplet fragments. As a result, this bonding model allows a linear structure, as shown in **Figure 1(A)**. In model (B), both E_{13}=R and Bi=R units still exist as triplets, so this bonding scheme contains one σ bond and one p-π bond (indicated by two dashed lines), plus one donor-acceptor π-bond because of coupling between the lone pair in Bi=R and the empty p orbital at the E_{13} atom (indicated by the arrow). Accordingly, this bonding pattern results in a bent structure, as shown in **Figure 1(B)**. The importance of the RE_{13}←BiR donor-acceptor interaction is emphasized, as it is essential for the stabilization of the nonlinear structure. These analyses are used to explain the geometrical structures of triply bonded RE_{13}≡BiR species in the following sections.

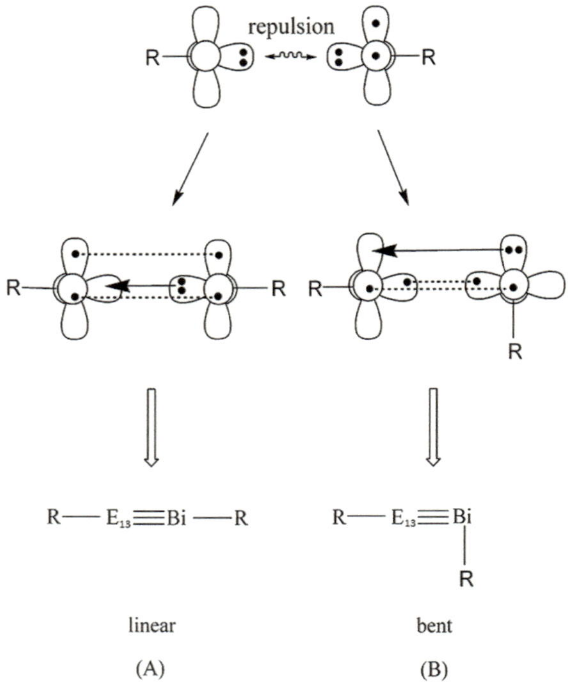

Figure 1. Two interaction models, A and B, in forming triply bonded RE_{13}≡BiR species.

3.2. Small ligands on substituted RE_{13}≡BiR

Small ligands, such as R = F, OH, H, CH_3, and SiH_3, are firstly chosen to study the geometries of the RE_{13}≡BiR (E_{13} = B, Al, Ga, In, and Tl) species. As mentioned in the Introduction, neither experimental nor theoretical results for the triply bonded RE_{13}≡BiR species are available to allow a definitive comparison. As a result, three DFT methods were used (i.e., M06-2X/Def2-

TZVP, B3PW91/Def2-TZVP, and B3LYP/LANL2DZ+dp) to examine their molecular properties. The selected geometrical parameters, natural charge densities (Q_{E13} and Q_{Bi}), binding energies (BE), and Wiberg bond order (BO) [63, 64] are shown in **Table 1** (RB≡BiR), **Table 2** (RAl≡BiR), **Table 3** (RGa≡BiR), **Table 4** (RIn≡BiR), and **Table 5** (RTl≡BiR).

Several important conclusions can be found in **Tables 1–5**, which are shown as follows:

R	F	OH	H	CH_3	SiH_3
B≡Bi (Å)	2.218	2.202	2.091	2.141	2.075
	−2.210	−2.199	−2.083	−2.137	−2.084
	[2.196]	[2.196]	[2.083]	[2.140]	[2.085]
∠R−B−Bi (°)	177.6	176	163.6	176.9	170.9
	−178.8	−175.9	−163.6	−176.9	−169.4
	[178.1]	[176.1]	[163.9]	[174.7]	[171.9]
∠B−Bi−R (°)	80.89	91.96	34.78	90.62	58.89
	−88.58	−90.21	−34.83	−90.26	−58.47
	[87.52]	[89.53]	[38.88]	[99.39]	[59.00]
∠R−B−Bi−R (°)	179.8	77.43	180	173.3	179.7
	−179.1	−75.70	−180.0	−173.5	−179.5
	[179.5]	[76.80]	[180.0]	[179.9]	[180.0]
Q_B (1)	0.1096	−0.0543	−0.3684	−0.3023	−0.5098
	−0.124	(−0.0372)	(−0.2262)	(−0.1390)	(−0.4100)
	[0.2303]	[0.0511]	[−0.1623]	[−0.01010]	[−0.4101]
Q_{Bi} (2)	0.4759	0.3569	0.2189	0.1881	0.1784
	−0.4975	−0.3431	−0.1331	−0.1602	−0.103
	[0.4983]	[0.3552]	[0.1870]	[0.1430]	[0.1520]

Notes: (1) The natural charge density on the central boron atom. (2) The natural charge density on the central bismuth atom. (3) BE = E (triplet state of R–B) + E (triplet state of R–Bi) – E(RB≡BiR). (4) Wiberg bond orders for the B–Bi bonds, see Ref. [18].

Table 1. Selected geometrical parameters, natural charge densities (Q_B and Q_{Bi}), binding energies (BE), and Wiberg bond orders (BO) of RB≡BiR at the M06-2X/Def2-TZVP, B3PW91/Def2-TZVP (in round bracket), and B3LYP/LANL2DZ+dp (in square bracket) levels.

R	F	OH	H	CH_3	SiH_3
Al≡Bi (Å)	2.604	2.606	2.439	2.529	2.501
	(2.601)	(2.601)	(2.463)	(2.539)	(2.532)
	[2.621]	[2.624]	[2.483]	[2.561]	[2.542]
∠R−Al−Bi (°)	176.8	173.4	170.7	177.7	172.5
	(175.4)	(172.9)	(168.1)	(177.6)	(170.4)
	[177.3]	[174.2]	[166.9]	[177.1]	[174.5]

R	F	OH	H	CH_3	SiH_3
∠Al−Bi−R (°)	83.16	84.20	48.09	92.52	62.57
	(84.59)	(85.35)	(49.85)	(92.93)	(61.82)
	[87.00]	[88.32]	[51.00]	[93.77]	[64.03]
∠R−Al−Bi−R (°)	180.0	177.3	180.0	179.9	179.9
	(180.0)	(176.0)	(180.0)	(179.6)	(179.8)
	[180.0]	178.0]	[180.0]	[179.6]	[180.0]
Q_{Al} (1)	0.5031	0.3942	0.1493	0.2692	0.1841
	(0.4904)	(0.3918)	(0.1417)	(0.2544)	(0.2145)
	[0.6664]	[0.4315]	[0.3786]	[0.2414]	[0.1517]
Q_{Bi} (2)	0.3947	0.2709	−0.05788	0.03761	−0.1384
	(0.3196)	(0.1834)	(−0.04954)	(0.02100)	(−0.07446)
	[0.3044]	[0.1982]	[0.03410]	[−0.05262]	[−0.1074]
BE (kcal mol^{-1}) (3)	22.61	20.28	50.55	38.69	53.41
	(30.36)	(31.77)	(85.64)	(63.54)	(57.96)
	[25.47]	[20.51]	[53.65]	[42.77]	[53.47]
Wiberg BO (4)	1.393	1.403	1.746	1.634	1.602
	(1.509)	(1.511)	(1.798)	(1.690)	(1.615)
	[1.521]	[1.516]	[1.787]	[1.706]	[1.653]

Notes: (1) The natural charge density on the central aluminum atom. (2) The natural charge density on the central bismuth atom. (3) BE = E (triplet state of R=Al) + E (triplet state of R=Bi) – E(RAl≡BiR). (4) Wiberg bond orders for the Al=Bi bonds, see Refs. [63, 64].

Table 2. Selected geometrical parameters, natural charge densities (Q_{Al} and Q_{Bi}), binding energies (BE), and Wiberg bond orders (BO) of RAl≡BiR at the M06-2X/Def2-TZVP, B3PW91/Def2-TZVP (in round bracket), and B3LYP/LANL2DZ +dp (in square bracket) levels.

R	F	OH	H	CH_3	SiH_3
Ga≡Bi (Å)	2.639	2.625	2.463	2.543	2.512
	(2.602)	(2.621)	(2.465)	(2.524)	(2.510)
	[2.632]	[2.629]	[2.487]	[2.550]	[2.520]
∠R−Ga−Bi (°)	179.7	175.0	166.5	178.9	178.7
	(178.3)	(173.3)	(166.2)	(177.8)	(177.6)
	[177.3]	[175.5]	[167.0]	[177.3]	[177.0]
∠Ga−Bi−R (°)	86.32	86.85	52.56	91.27	65.56
	(88.49)	(88.52)	(56.24)	(92.86)	(66.28)
	[88.18]	[90.75]	[59.49]	[93.37]	[69.82]
∠R−Ga−Bi−R (°)	179.5	157.1	180.0	179.2	175.8
	(180.0)	(159.8)	(180.0)	(178.8)	(179.9)

http://dx.doi.org/10.5772/67220

R	F	OH	H	CH_3	SiH_3
	[180.0]	[158.0]	[180.0]	[179.3]	[180.0]
Q_{Ga} (1)	0.6246	0.5224	0.2127	0.2266	0.1845
	(0.5012)	(0.3464)	(0.1135)	(0.2356)	(0.1507)
	[0.5700]	[0.3813]	[0.3031]	[0.1984]	[0.07925]
Q_{Bi} (2)	0.3696	0.2435	−0.1367	0.03356	−0.2002
	(0.3574)	(0.2743)	(−0.04212)	(0.05503)	(−0.05512)
	[0.4000]	[0.2395]	[0.06523]	[−0.01620]	[−0.08931]
BE (kcal mol^{-1}) (3)	18.03	16.41	45.28	46.24	46.10
	(22.92)	(26.32)	(79.05)	(60.77)	(49.96)
	[20.80]	[16.61]	[49.49]	[40.12]	[48.61]
Wiberg BO (4)	1.286	1.335	1.718	1.578	1.653
	(1.382)	(1.393)	(1.787)	(1.633)	(1.646)
	[1.403]	[1.431]	[1.758]	[1.656]	[1.673]

Notes: (1) The natural charge density on the central gallium atom. (2) The natural charge density on the central bismuth atom. (3) BE = E (triplet state of R=Ga) + E (triplet state of R=Bi) – E(RGa≡BiR). (4) Wiberg bond orders for the Ga=Bi bonds, see Refs. [63, 64].

Table 3. Selected geometrical parameters, natural charge densities (Q_{Ga} and Q_{Bi}), binding energies (BE), and Wiberg bond orders (BO) of RGa≡BiR at the M06-2X/Def2-TZVP, B3PW91/Def2-TZVP (in round bracket), and B3LYP/LANL2DZ +dp (in square bracket) levels.

R	F	OH	H	CH_3	SiH_3
In≡Bi (Å)	2.804	2.790	2.659	2.696	2.667
	(2.802)	(2.700)	(2.691)	(2.719)	(2.692)
	[2.790]	[2.795]	[2.673]	[2.712]	[2.683]
∠R—In—Bi (°)	179.6	173.9	168.3	179.1	173.8
	(178.0)	(172.2)	(175.2)	(177.3)	(174.3)
	[177.0]	[174.5]	[174.1]	[177.6]	[174.7]
∠In—Bi—R (°)	84.16	85.14	67.00	92.20	70.37
	(87.83)	(89.71)	(77.82)	(94.83)	(75.05)
	[87.43]	[90.79]	[74.35]	[94.16]	[74.37]
∠R—In—Bi—R (°)	179.9	176.1	180.0	178.3	178.9
	(180.0)	(177.3)	(180.0)	(179.1)	(177.5)
	[180.0]	[176.5]	[180.0]	[179.4]	[179.8]
Q_{In} (1)	0.7021	0.6352	0.3755	0.3297	0.3650
	(0.5692)	(0.4640)	(0.2561)	(0.3403)	(0.2872)
	[0.7571]	[0.5053]	[0.4055]	[0.3044]	[0.1756]
Q_{Bi} (2)	0.3973	0.2452	−0.2474	−0.01187	−0.2599

R	F	OH	H	CH_3	SiH_3
	(0.4000)	(0.2511)	(−0.08703)	(0.04410)	(−0.08023)
	[0.3468]	[0.2141]	[0.02620]	[−0.05735]	[−0.1365]
BE (kcal mol^{-1}) (3)	14.66	13.17	39.19	36.87	39.28
	(15.06)	(12.88)	(42.01)	(35.00)	(40.94)
	[18.80]	[13.70]	[44.04]	[35.54]	[41.83]
Wiberg BO (4)	1.312	1.403	1.590	1.543	1.553
	(1.308)	(1.334)	(1.601)	(1.539)	(1.546)
	[1.323]	[1.336]	[1.615]	[1.548]	[1.549]

Notes: (1) The natural charge density on the central indium atom. (2) The natural charge density on the central bismuth atom. (3) BE = E (triplet state of R–In) + E (triplet state of R–Bi) – E(RIn≡BiR). (4) Wiberg bond orders for the In–Bi bonds, see Refs. [63, 64].

Table 4. Selected geometrical parameters, natural charge densities (Q_{In} and Q_{Bi}), binding energies (BE), and Wiberg bond orders (BO) of RIn≡BiR at the M06-2X/Def2-TZVP, B3PW91/Def2-TZVP (in round bracket), and B3LYP/LANL2DZ+dp (in square bracket) levels.

R	F	OH	H	CH_3	SiH_3
Tl≡Bi (Å)	2.859	2.843	2.713	2.742	2.707
	(2.812)	(2.803)	(2.698)	(2.725)	(2.705)
	[2.819]	[2.822]	[2.679]	[2.713]	[2.682]
∠R—Tl—Bi (°)	175.9	173.5	175.7	179.6	174.4
	(178.8)	(172.5)	(176.6)	(178.5)	(174.7)
	[177.2]	[174.4]	[176.3]	[178.5]	[175.9]
∠Tl—Bi—R (°)	81.34	86.72	78.87	91.91	76.03
	(87.73)	(89.82)	(79.00)	(93.54)	(76.50)
	[87.92]	[92.34]	[78.86]	[93.25]	[80.00]
∠R—Tl—Bi—R (°)	180.0	143.7	180.0	172.1	178.9
	(179.9)	(132.9)	(179.9)	(178.6)	(179.9)
	[180.0]	[130.6]	[180.0]	[180.0]	[179.2]
Q_{Tl} (1)	0.6481	0.5672	0.3014	0.4162	0.2665
	(0.6614)	(0.5879)	(0.2284)	(0.2746)	(0.3510)
	[0.7100]	[0.4812]	[0.3536]	[0.2734]	[0.1601]
Q_{Bi} (2)	0.4752	0.3214	−0.1836	0.008121	−0.2245
	(0.3615)	(0.2201)	(−0.05213)	(−0.1282)	(−0.09637)
	[0.3854]	[0.2455]	[0.04813]	[−0.03131]	[−0.1282]
BE (kcal mol^{-1}) (5)	7.90	6.61	30.74	26.94	30.56
	(2.98)	(8.17)	(48.12)	(35.39)	(29.13)

http://dx.doi.org/10.5772/67220

R	F	OH	H	CH_3	SiH_3
	[11.34]	[7.38]	[37.14]	[31.09]	[35.03]
Wiberg BO [(6)]	1.000	1.132	1.652	1.448	1.792
	(1.121)	(1.254)	(1.765)	(1.514)	(1.824)
	[1.112]	[1.212]	[1.756]	[1.568]	[1.652]

Notes: (1) The natural charge density on the central thallium atom. (2) The natural charge density on the central bismuth atom. (3) BE = E (triplet state of R–Tl) + E (triplet state of R–Bi) – E(RTl≡BiR). (4) Wiberg bond orders for the Tl≡Bi bonds, see Refs. [63, 64].

Table 5. Selected geometrical parameters, natural charge densities (Q_{Tl} and Q_{Bi}), binding energies (BE), and Wiberg bond orders (BO) of RTl≡BiR at the M06-2X/Def2-TZVP, B3PW91/Def2-TZVP (in round bracket), and B3LYP/LANL2DZ+dp (in square bracket) levels.

1. As can be seen from **Tables 1–5**, the geometrical parameters of the RE_{13}≡BiR triple bond species are quite analogous at the three levels employed. For instance, the predicted triple bond length for small ligands on RB≡BiR is 2.075–2.218 Å (M06-2X/Def2-TZVP), 2.083–2.210 Å (B3PW91/Def2-TZVP), and 2.083–2.196 Å (B3LYP/LANL2DZ+dp). For RAl≡BiR, the Al≡Bi triple bond length for small ligands is predicted to be 2.439–2.606 Å (M06-2X/Def2-TZVP), 2.463–2.601 Å (B3PW91/Def2-TZVP), and 2.483–2.624 Å (B3LYP/LANL2DZ+dp). For RGa≡BiR, the Ga≡Bi triple bond length for small ligands is predicted to be 2.463–2.639 Å (M06-2X/Def2-TZVP), 2.465–2.621 Å (B3PW91/Def2-TZVP), and 2.487–2.632 Å (B3LYP/LANL2DZ+dp). For RIn≡BiR, the In≡Bi triple bond length for small ligands is predicted to be 2.659–2.804 Å (M06-2X/Def2-TZVP), 2.691–2.802 Å (B3PW91/Def2-TZVP), and 2.673–2.795 Å (B3LYP/LANL2DZ+dp). For RTl≡BiR, the Tl≡Bi triple bond length for small ligands is predicted to be 2.707–2.859 Å (M06-2X/Def2-TZVP), 2.698–2.812 Å (B3PW91/Def2-TZVP), and 2.679–2.822 Å (B3LYP/LANL2DZ+dp).

2. It is apparent from **Tables 1–5** that an acute bond angle $\angle E_{13}$≡Bi–R (close to 90°) in the triply bonded molecule RE_{13}≡BiR is favored. The reason for this can be attributed to the "orbital nonhybridization effect," also known as the "inert *s*-pair effect" [65–68], as discussed previously. Accordingly, these phenomena strongly indicate that mode (B) (**Figure 1**) is preferred in the RE≡BiR molecule, for which bent geometry is favored.

3. The Wiberg bond orders (WBOs) [63, 64] on the substituted RE_{13}≡BiR compounds are also given in **Tables 1–5**. For all the triply bonded RE_{13}≡BiR molecules with small substituents, their WBOs were computed to be less than 2.0, except for the cases of HB≡BiH and (SiH_3)B≡Bi(SiH_3). These WBO values imply that the bonding structure of RE_{13}≡BiR may be due to the resonance structures, [I] and [II]. That is to say, the E_{13}≡Bi bond could be either double or triple bonds. From **Tables 1–5**, it seems that the resonance structure [II] prevails for the small ligands on the substituted RE_{13}≡BiR species studied in this work. Indeed, since it is known that the electronegativities decrease in the order B (2.051) > Bi (2.01) > Tl (1.789) > Ga (1.756) > In (1.656) > Al (1.613) [69], the bonding mode of RE_{13}≡BiR should prefer to adopt resonance structure [II] (**Scheme 2**).

With regard to the stability of $RE_{13}BiR$, the results of theoretical calculations on the energy surface of the model $RE_{13}BiR$ (R = F, OH, H, CH_3, and SiH_3) system are depicted in **Figures 2–6**.

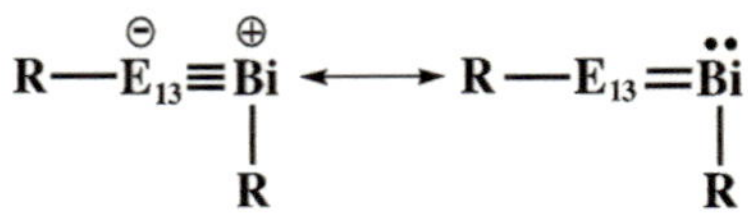

[I] [II]

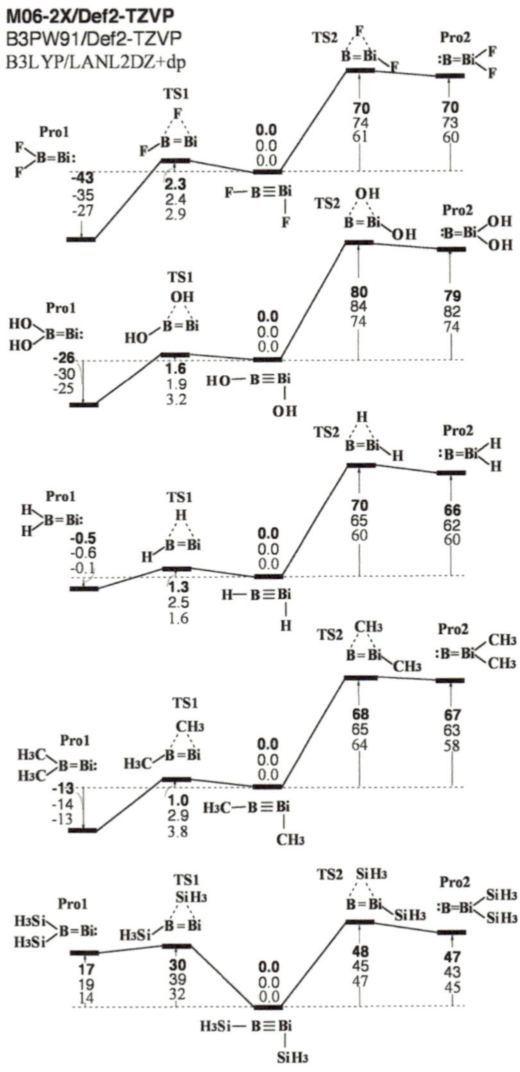

Figure 2. Relative Gibbs free energy surfaces for RB≡BiR (R = F, OH, H, CH_3, and SiH_3). Energies are in kcal/mol, calculated at M06-2X/Def2-TZVP, B3PW91/Def2-TZVP, and B3LYP/LANL2DZ+dp levels of theory. For details see the text and **Table 1**.

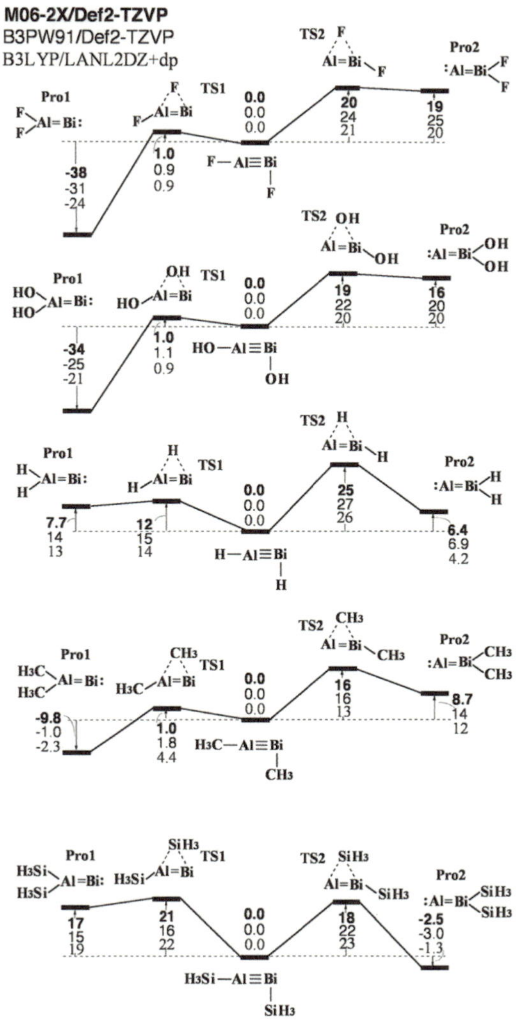

Figure 3. Relative Gibbs free energy surfaces for RAl≡BiR (R = F, OH, H, CH_3, and SiH_3). Energies are in kcal/mol, calculated at M06-2X/Def2-TZVP, B3PW91/Def2-TZVP, and B3LYP/LANL2DZ+dp levels of theory. For details see the text and **Table 2**.

This system exhibits a number of stationary points, including local minima that correspond to RE_{13}≡BiR, R_2E_{13}=Bi:, :E_{13}=BiR_2, and the saddle points connecting them. The transition structures that separate the three stable molecular forms involve a successive unimolecular 1,2-shift TS1 (from RE_{13}≡BiR to R_2E_{13}=Bi:) and a 1,2-shift TS2 (from RE_{13}≡BiR to :E_{13}=BiR_2). As shown in **Figures 1–5**, these theoretical studies using the M06-2X, B3PW91, and B3LYP levels show that the RE_{13}≡BiR species are local minima on the singlet potential energy surface, but they are neither kinetically nor thermodynamically stable for small substituents, except for the case of (SiH_3)B≡Bi(SiH_3). As a result, these triply bonded structures RE_{13}≡BiR seem to be unstable on the singlet energy surface and undergo unimolecular rearrangement to the doubly bonded isomer. In brief, these triply bonded molecules (RE_{13}≡BiR) possessing the small substituents are predicted to be a kinetically unstable isomer, so these could not be isolated in a matrix or even as transient intermediates.

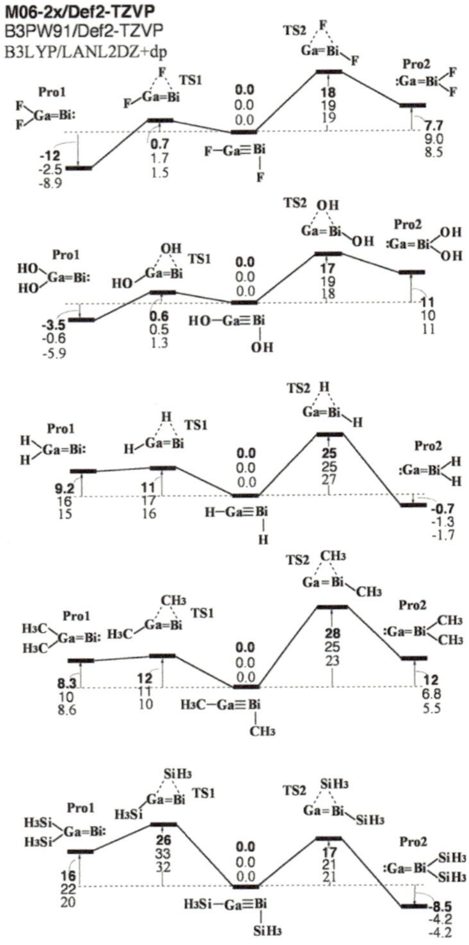

Figure 4. Relative Gibbs free energy surfaces for RGa≡BiR (R = F, OH, H, CH_3, and SiH_3). Energies are in kcal/mol, calculated at M06-2X/Def2-TZVP, B3PW91/Def2-TZVP, and B3LYP/LANL2DZ+dp levels of theory. For details see the text and **Table 3**.

3.3. Large ligands on substituted $R'E_{13}$≡BiR'

According to the above conclusions for the cases of small substituents, it is necessary to determine whether bulky substituents can destabilize R_2E_{13}=Bi: and :E_{13}=BiR_2 relative to RE_{13}≡BiR (E_{13} = B, Al, Ga, In, and Tl), due to severe steric overcrowding. From **Figure 7**, it is easily anticipated that the presence of extremely bulky substituents at both ends of the RE_{13}≡BiR compounds protects its triple bond from intermolecular reactions, such as polymerization. In order to examine the effect of bulky substituents, the structures of $R'E_{13}$≡BiR' optimized for R' = Tbt, Ar*, SiMe(Si*t*Bu₃)₂, and Si*i*PrDis₂ (Scheme 1) at the B3LYP/LANL2DZ +dp level. Selected geometrical parameters, natural charge densities on the central group 13 elements and bismuth (Q_{E13} and Q_{Bi}), binding energies (BE), and Wiberg bond order (BO) [69, 70] are summarized in **Tables 6–10**.

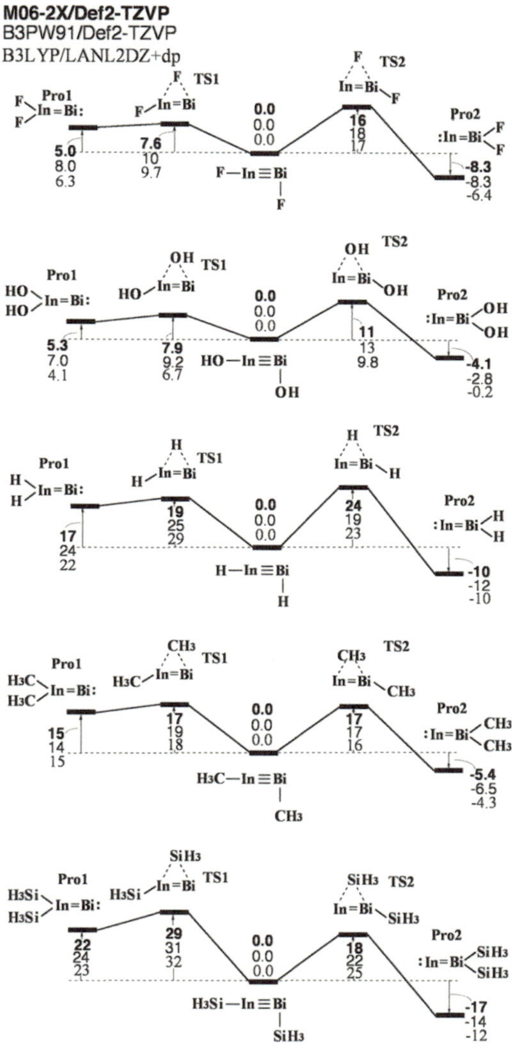

Figure 5. Relative Gibbs free energy surfaces for RIn≡BiR (R = F, OH, H, CH_3, and SiH_3). Energies are in kcal/mol, calculated at M06-2X/Def2-TZVP, B3PW91/Def2-TZVP, and B3LYP/LANL2DZ+dp levels of theory. For details see the text and **Table 4**.

The computational results given in **Tables 6–10** estimate that the E_{13}≡Bi triple bond distances (Å) are about 2.117–2.230 (E_{13} = B), 2.461–2.562 (E_{13} = Al), 2.576–2.580(E_{13} = Ga), 2.615–2.779 (E_{13} = In), and 2.789–2.833 (E_{13} = Tl), respectively. Again, these theoretically predicted values are much shorter than the available experimentally determined E_{13}–Bi single bond lengths

[36, 70–72]. This strongly implies that the central group 13 element (E_{13}) and bismuth in the R'E_{13}≡BiR' (R' = Tbt, Ar*, SiMe(Si*t*Bu$_3$)$_2$, and Si*i*PrDis$_2$) species are triply bonded. Indeed, as shown in **Tables 6–10**, the R'E_{13}≡BiR' molecules accompanied by bulky ligands can effectively produce the triply bonded species. That is, the WBOs in **Tables 6–10** (with larger ligands) are apparently larger than those in **Tables 1–5** (with smaller ligands). Additionally, from **Tables 6–10**, the central E_{13}≡Bi bond lengths calculated for R' = Tbt and Ar* are an average 0.095Å longer than those calculated for R' = SiMe(Si*t*Bu$_3$)$_2$ and Si*i*PrDis$_2$, respectively. The reason for these differences is that the Tbt and Ar* groups are electronegative, but the SiMe (Si*t*Bu$_3$)$_2$ and Si*i*PrDis$_2$ ligands are electropositive. Further, the short length of the E_{13}≡Bi bond in the R'E≡BiR' species can be understood by noting that both SiMe(Si*t*Bu$_3$)$_2$ and Si*i*PrDis$_2$ are more electropositive than the small substituents, as mentioned earlier.

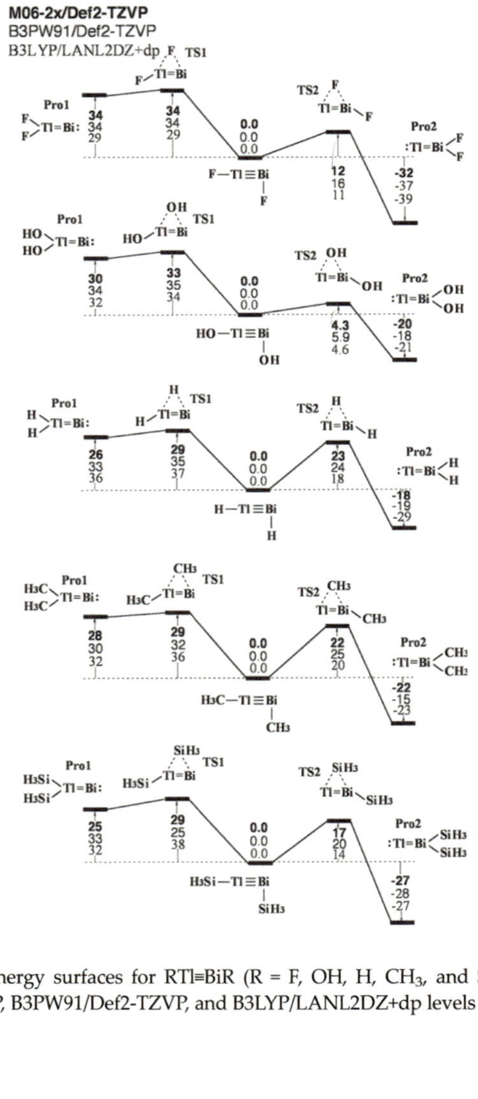

Figure 6. Relative Gibbs free energy surfaces for RTl≡BiR (R = F, OH, H, CH_3, and SiH_3). Energies are in kcal/mol, calculated at M06-2X/Def2-TZVP, B3PW91/Def2-TZVP, and B3LYP/LANL2DZ+dp levels of theory. For details see the text and **Table 5**.

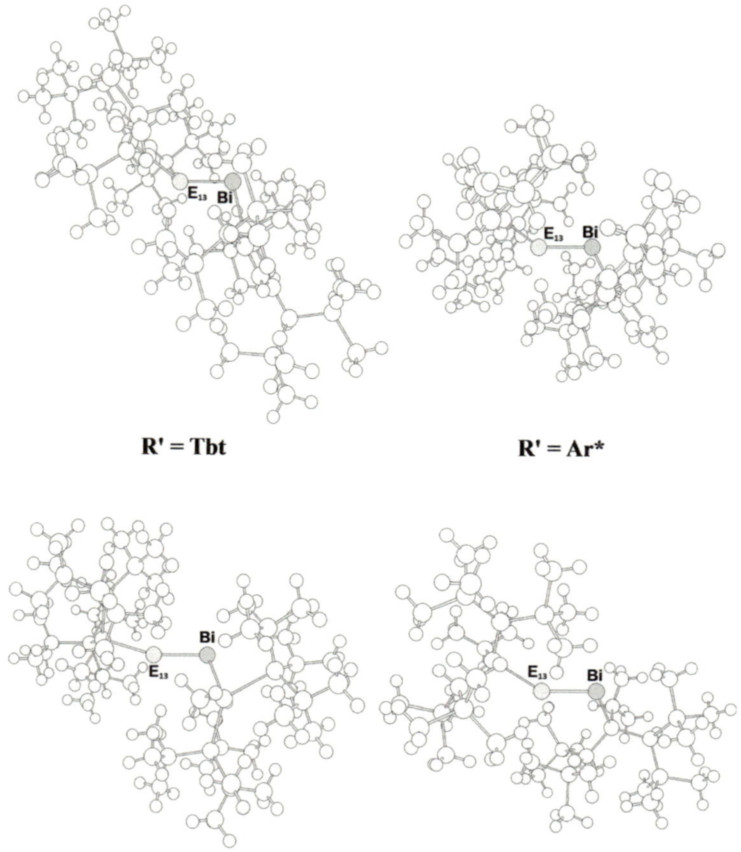

Figure 7. The optimized structures of R'E_{13}≡BiR' (E_{13} = B, Al, Ga, In, and Tl; R' = Tbt, Ar*, SiMe(Si*t*Bu_3)$_2$, and Si*i*PrDis_2) at the B3LYP/LANL2DZ+dp level of theory. For details see the text and **Tables 6–10**.

Similar to the small ligands, these DFT results demonstrate that all the R'E_{13}BiR' molecules that possess bulky substituents (R') adopt a bent geometry, as illustrated in **Figure** 7. Our theoretical computations show that model (B), given in **Figure 1**, still predominates and can be used to interpret the geometries of the R'E_{13}≡BiR' systems that bear bulky substituents.

As shown in **Tables 6–10**, the R'E_{13}≡BiR' molecules can be separated into two fragments in solution, when the substituent R' becomes bulkier. The BE that is essential to break the central E_{13}≡Bi bond was computed to be at least > 32 kcal/mol for R' = Tbt, Ar*, SiMe(Si*t*Bu_3)$_2$, and Si*i*PrDis_2, for the B3LYP/LANL2DZ+dp method, as given in **Tables 6–10**. These BE values show that the central E_{13} and bismuth elements are strongly bonded and R'E_{13}≡BiR' molecules that contain bulky substituents do not dissociate in solution. Namely, the larger the dissociation energy of the E_{13}≡Bi bond, the shorter and stronger the E_{13}≡Bi triple bond.

As predicted previously, bulky groups destabilize the 1,2-R' migrated isomers because they crowd around one end of the central E_{13}≡Bi bond. As a consequence, the bulky substituents (R') can prevent the isomerization of R'E_{13}≡BiR' compounds, as outlined in **Scheme 3** and

Tables 6–10. The B3LYP/LANL2DZ+dp calculations indicate that the R'E$_{13}$≡BiR' species with Tbt, Ar*, SiMe(Si*t*Bu$_3$)$_2$, and Si*i*PrDis$_2$ substituents (ΔH_1 and ΔH_2) are at least 56 kcal/mol more stable than the 1,2-R' shifted isomers, respectively. These theoretical results suggest that both doubly bonded R'$_2$E$_{13}$=Bi: and :E$_{13}$=BiR'$_2$ isomers are kinetically and thermodynamically unstable, so they rearrange spontaneously to the global minimum R'E$_{13}$≡BiR' triply bonded molecules, provided that significantly bulky groups are employed.

Theoretical values from the natural bond orbital (NBO) [63, 64] and natural resonance theory (NRT) [73–75] analyses of the R'E$_{13}$≡BiR' molecules, computed at the B3LYP/LANL2DZ+dp level of theory, are summarized in **Table 11** (E$_{13}$ = B), **Table 12** (E$_{13}$ = Al), **Table 13** (E$_{13}$ = Ga), **Table 14** (E$_{13}$ = In), and **Table 15** (E$_{13}$ = Tl).

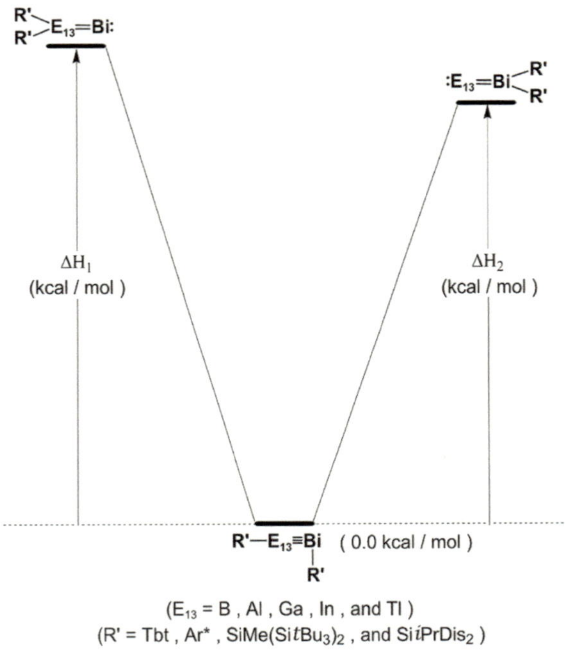

R'	Tbt	Ar*	SiMe(Si*t*Bu$_3$)$_2$	Si*i*PrDis$_2$
B≡Bi (Å)	2.230	2.214	2.117	2.131
∠R'–B–Bi (°)	177.3	110.3	112.9	113.6
∠B–Bi–R' (°)	115.6	115.5	112.5	114.3
∠R'–B–Bi–R' (°)	173.7	172.3	170.4	175.3
Q_B (1)	−0.4310	−0.1711	−0.3251	−0.4742
Q_{Bi} (2)	0.2915	0.3004	0.1426	0.1071
BE (kcal mol^{-1}) (3)	37.58	41.68	36.25	51.07

R'	Tbt	Ar*	SiMe(Si*t*Bu$_3$)$_2$	Si*i*PrDis$_2$
Wiberg BO $^{(4)}$	2.385	2.249	2.640	2.701
ΔH_1 (kcal mol^{-1}) $^{(5)}$	62.6	61.04	76.52	77.24
ΔH_2 (kcal mol^{-1}) $^{(6)}$	98.51	88.38	68.72	78.13

Notes: (1) The natural charge density on the central boron atom. (2) The natural charge density on the central bismuth atom. (3) BE = E (triplet state of B–R') + E (triplet state of Bi–R') – E(R'B≡BiR'). (4) Wiberg Bond Orders for the B–Bi bond, see Refs. [63, 64]. (5) ΔH_1 = E(:B=BiR'$_2$) – E(R'B≡BiR'); see Scheme 3. (6) ΔH_2 = E(R'$_2$B=Bi:) – E(R'B≡BiR'); see **Scheme 3**

Table 6. Geometrical parameters, nature charge densities (Q_B and Q_{Bi}), binding energies (BE), and Wiberg bond order (BO) of R'B≡BiR' at the B3LYP/LANL2DZ+dp level of theory. Also see **Figure 7**.

R'	Tbt	Ar*	SiMe(Si*t*Bu$_3$)$_2$	Si*i*PrDis$_2$
Al≡Bi (Å)	2.562	2.561	2.463	2.461
∠R'—Al—Bi (°)	178.0	113.6	115.4	113.3
∠Al—Bi—R' (°)	113.4	115.5	112.2	109.0
∠R'—Al—Bi—R' (°)	167.1	165.8	173.2	174.7
Q_{Al} $^{(1)}$	0.4171	0.4111	0.2173	0.1585
Q_{Bi} $^{(2)}$	0.0773	0.2208	−0.1031	−0.1862
BE (kcal mol^{-1}) $^{(3)}$	38.62	66.78	33.65	31.76
Wiberg BO $^{(4)}$	2.092	2.023	2.204	2.259
ΔH_1 (kcal mol^{-1}) $^{(5)}$	64.52	63.77	77.71	73.68
ΔH_2 (kcal mol^{-1}) $^{(6)}$	57.43	67.81	71.86	76.62

Notes: (1) The natural charge density on the central aluminum atom. (2) The natural charge density on the central bismuth atom. (3) BE = E (triplet state of Al–R') + E (triplet state of Bi–R') – E(R'Al≡BiR'). (4) Wiberg Bond Orders for the Al≡Bi bond, see Refs. [63, 64]. (5) ΔH_1 = E(:Al=BiR'$_2$) – E(R'Al≡BiR'); see Scheme 3. (6) ΔH_2 = E(R'$_2$Al=Bi:) – E(R'Al≡BiR'); see **Scheme 3**.

Table 7. Geometrical parameters, nature charge densities (Q_{Al} and Q_{Bi}), binding energies (BE), and Wiberg bond order (BO) of R'Al≡BiR' at the B3LYP/LANL2DZ+dp level of theory. Also see **Figure 7**.

All the NBO values listed in **Tables 11–15** demonstrate that there exists a weak triple bond, or perhaps a bond between a double and a triple, in the ethyne-like R'E_{13}≡BiR' molecule. For instance, the B3LYP/LANL2DZ+dp data for the NBO [63, 64] analyses of the B≡Bi bonding in SiMe(Si*t*Bu$_3$)$_2$–B≡Bi–SiMe(Si*t*Bu$_3$)$_2$, which shows that NBO(B≡Bi) = 0.615(2s2p$^{52.48}$)B + 0.789 (6s6p$^{19.73}$)Bi, strongly suggests that the predominant bonding interaction between the B–SiMe (Si*t*Bu$_3$)$_2$ and the Bi–SiMe(Si*t*Bu$_3$)$_2$ fragments originates from 2p(B) ← 6p(Bi) donation. In other words, boron's electron deficiency and π bond polarity are partially balanced by the donation of the bismuth lone pair into the empty boron p orbital. This, in turn, forms a hybrid π bond. Again, the polarization analyses using the NBO model indicate the presence of the B≡Bi π bonding orbital, 38% of which is composed of natural boron orbitals and 62% of natural bismuth orbitals. There is supporting evidence in **Table 11** that reveals that the B≡Bi triple bond in SiMe(Si*t*Bu$_3$)$_2$–B≡Bi–SiMe(Si*t*Bu$_3$)$_2$ has a shorter single bond character (5.8%) and a

R'	Tbt	Ar*	SiMe(Si*t*Bu$_3$)$_2$	Si*i*PrDis$_2$
Ga≡Bi (Å)	2.578	2.576	2.580	2.579
∠R'−Ga−Bi (°)	178.1	113.4	115.2	112.1
∠Ga−Bi−R' (°)	113.4	115.7	112.0	110.1
∠R'−Ga−Bi−R' (°)	167.9	164.1	175.4	178.5
Q_{Ga} (1)	0.240	0.196	0.069	0.012
Q_{Bi} (2)	0.120	0.261	−0.055	−0.140
BE (kcal mol^{-1}) (3)	43.73	39.28	35.54	32.92
Wiberg BO (4)	2.091	2.181	2.262	2.313
ΔH_1 (kcal mol^{-1}) (5)	68.10	69.08	61.74	58.83
ΔH_2 (kcal mol^{-1}) (6)	78.07	71.28	77.43	64.13

Notes: (1) The natural charge density on the central gallium atom. (2) The natural charge density on the central bismuth atom. (3) BE = E (triplet state of Ga=R') + E (triplet state of Bi=R') – E(R'Ga≡BiR'). (4) Wiberg bond orders for the Ga=Bi bond, see Refs. [63, 64]. (5) ΔH_1 = E(:Ga = BiR'$_2$) – E(R'Ga≡BiR'); see Scheme 3. (6) ΔH_2 = E(R'$_2$Ga=Bi:) – E(R'Ga≡BiR'); see **Scheme 3**.

Table 8. Geometrical parameters, nature charge densities (Q_{Ga} and Q_{Bi}), binding energies (BE), and Wiberg bond order (BO) of R'Ga≡BiR' at the B3LYP/LANL2DZ+dp level of theory. Also see **Figure 7**.

R'	Tbt	Ar*	SiMe(Si*t*Bu$_3$)$_2$	Si*i*PrDis$_2$
In≡Bi (Å)	2.737	2.779	2.615	2.678
∠R'−In−Bi (°)	178.7	111.7	110.0	110.9
∠In−Bi−R' (°)	112.5	113.0	110.6	111.7
∠R'−In−Bi−R' (°)	170.7	174.6	164.3	162.0
Q_{In} (1)	0.299	0.345	0.179	0.101
Q_{Bi} (2)	0.066	0.293	−0.126	−0.132
BE (kcal mol^{-1}) (3)	63.45	45.97	36.20	37.05
Wiberg BO (4)	2.052	2.153	2.211	2.304
ΔH_1 (kcal mol^{-1}) (5)	64.06	60.17	55.72	62.99
ΔH_2 (kcal mol^{-1}) (6)	79.38	61.44	56.03	67.61

Notes: (1) The natural charge density on the central indium atom. (2) The natural charge density on the central bismuth atom. (3) BE = E (triplet state of In=R') + E (triplet state of Bi=R') – E(R'In≡BiR'). (4) Wiberg Bond Orders for the In=Bi bond, see Refs. [63, 64]. (5) ΔH_1 = E(:In = BiR'$_2$) – E(R'In≡BiR'); see Scheme 3. (6) ΔH_2 = E(R'$_2$In=Bi:) – E(R'In≡BiR'); see **Scheme 3.**

Table 9. Geometrical parameters, nature charge densities (Q_{In} and Q_{Bi}), binding energies (BE), and Wiberg bond order (BO) of R'In≡BiR' at the B3LYP/LANL2DZ+dp level of theory. Also see **Figure 7**.

shorter triple bond character (40.1%) but a larger double bond character (54.1%), because the covalent part of the NRT bond order (1.49) is shorter than its ionic part (0.78). The same can also be said of the other three R'B≡BiR' molecules, as shown in **Table 11** as well as other

http://dx.doi.org/10.5772/67220

R'E$_{13}$≡BiR' compounds represented in **Tables 12–15**. These theoretical evidences strongly suggest that these R'E$_{13}$≡BiR' species have a weak triple bond.

R'	Tbt	Ar*	SiMe(Si*t*Bu$_3$)$_2$	Si*i*PrDis$_2$
Tl≡Bi (Å)	2.816	2.833	2.820	2.789
∠R'−Tl−Bi (°)	148.2	146.7	164.4	152.4
∠Tl−Bi−R' (°)	116.5	121.4	114.1	125.1
∠R'−Tl−Bi−R (°)	169.5	171.4	176.3	175.8
Q_{Tl} (1)	0.3323	0.4221	0.1504	0.1282
Q_{Bi} (2)	−0.0921	−0.0652	−0.3925	−0.3491
BE (kcal mol^{-1}) (3)	41.66	37.59	36.66	51.07
Wiberg BO (4)	2.081	2.052	2.105	2.128
ΔH_1 (kcal mol^{-1}) (5)	68.66	64.52	63.67	72.12
ΔH_2 (kcal mol^{-1}) (6)	61.54	58.19	68.73	64.61

Notes: (1) The natural charge density on the central thallium atom. (2) The natural charge density on the central bismuth atom. (3) BE = E (triplet state of Tl–R') + E (triplet state of Bi–R') – E(R'Tl≡BiR'). (4) Wiberg bond orders for the Tl–Bi bond, see Refs. [63, 64]. (5) ΔH_1 = E(:Tl = BiR'$_2$) – E(R'Tl≡BiR'); see Scheme 3. (6) ΔH_2 = E(R'$_2$Tl = Bi:) – E(R'Tl≡BiR'); see **Scheme 3**.

Table 10. Geometrical parameters, nature charge densities (Q_{Tl} and Q_{Bi}), binding energies (BE), and Wiberg bond order (BO) of R'Tl≡BiR' at the B3LYP/LANL2DZ+dp level of theory. Also see **Figure 7**.

R′B≡BiR′	WBI	NBO analysis			NRT analysis	
		Occupancy	Hybridization	Polarization	Total/covalent/ionic	Resonance weight
R′=Tbt	2.39	σ = 1.95	σ: 0.7870 B (sp$^{0.90}$) + 0.6170 Bi (sp$^{9.90}$)	61.93% (B)	2.01/1.24/0.77	B–Bi: 12.58%
				38.07% (Bi)		B=Bi: 53.81%
		π = 1.93	π: 0.5938 B (sp$^{1.00}$) + 0.8046 Bi (sp$^{1.00}$)	35.26% (B)		B≡Bi: 33.61%
				64.74% (Bi)		
R′ = Ar*	2.25	σ = 1.94	σ: 0.8058 B (sp$^{0.74}$) + 0.5922 Bi (sp$^{19.46}$)	64.93% (B)	1.95/1.30/0.65	B–Bi: 6.78%
				35.07% (Bi)		B=Bi: 62.97%
		π = 1.90	π: 0.5587 B (sp$^{99.99}$) + 0.8294 Bi (sp$^{64.69}$)	31.22% (B)		B≡Bi: 30.25%
				68.78% (Bi)		
R′ = SiMe (Si*t*Bu$_3$)$_2$	2.64	σ = 1.96	σ: 0.7812 B (sp$^{0.96}$) + 0.6242 Bi (sp$^{10.68}$)	61.03% (B)	2.27/1.49/0.78	B–Bi: 5.77%
				38.97% (Bi)		B=Bi: 54.1%

R'B≡BiR'	**WBI**	**NBO analysis**			**NRT analysis**	
		Occupancy	**Hybridization**	**Polarization**	**Total/covalent/ionic**	**Resonance weight**
		π = 1.89	π: 0.6146 B ($sp^{62.48}$) + 0.7889 Bi ($sp^{19.73}$)	37.77% (B) 62.23% (Bi)		B≡Bi: 40.11%
R' = Si*i*PrDis$_2$	2.7	σ = 1.83	σ: 0.6502 B ($sp^{4.29}$) + 0.7598 Bi ($sp^{1.07}$)	42.27% (B) 57.73% (Bi)	2.31/1.52/0.79	B=Bi: 6.01% B=Bi: 54.39%
		π = 1.80	π: 0.5606 B ($sp^{1.73}$) + 0.8281 Bi ($sp^{4.99}$)	31.43% (B) 68.57% (Bi)		B≡Bi: 39.96%

(1) The Wiberg bond index (WBI) for the B=Bi bond and occupancy of the corresponding σ and π bonding NBO: see Refs. [63, 64], and (2) the natural resonance theory (NRT): see Refs. [73–75].

Table 11. Selected results for the natural bond orbital (NBO) and natural resonance theory (NRT) analyses of R'B≡BiR' compounds that have small substituents, at the B3LYP/LANL2DZ+dp level of theory [1–8, 76–80].

R'Al≡BiR'	**WBI**	**NBO analysis**			**NRT analysis**	
		Occupancy	**Hybridization**	**Polarization**	**Total/covalent/ionic**	**Resonance weight**
R' = Tbt	2.09	σ = 1.98	σ: 0.7538 Al ($sp^{0.15}$) + 0.6571 Bi ($sp^{22.99}$)	56.83% (Al) 43.17% (Bi)	2.12/1.10/1.02	Al=Bi: 12.76% Al=Bi: 75.36%
		π = 1.93	π: 0.4709 Al ($sp^{1.00}$) + 0.8822 Bi ($sp^{1.00}$)	22.17% (Al) 77.83% (Bi)		Al≡Bi: 11.88%
R' = Ar*	2.02	σ = 1.84	σ: 0.7806 Al ($sp^{0.15}$) + 0.6250 Bi ($sp^{28.77}$)	60.93% (Al) 39.07% (Bi)	2.07/1.01/1.06	Al=Bi: 19.33% Al=Bi: 74.20%
		π = 1.94	π: 0.4960 Al ($sp^{46.09}$) + 0.8673 Bi ($sp^{15.43}$)	24.60% (Al) 75.40% (Bi)		Al≡Bi: 6.47%
R' = SiMe (Si*t*Bu$_3$)$_2$	2.2	σ = 1.96	σ: 0.7169 Al ($sp^{0.96}$) + 0.6971 Bi ($sp^{21.26}$)	24.70% (Al) 75.30% (Bi)	2.24/1.38/0.86	Al=Bi: 11.69% Al=Bi: 84.51%
		π = 1.89	π: 0.8678 Al ($sp^{19.21}$) + 0.4970 Bi ($sp^{16.37}$)	37.77% (Al) 62.23% (Bi)		Al≡Bi: 3.80%
R' = Si*i*PrDis$_2$	2.26	σ = 1.86	σ: 0.7184 Al ($sp^{0.93}$) + 0.6956 Bi ($sp^{29.72}$)	51.61% (Al) 48.39% (Bi)	1.91/1.35/0.60	Al=Bi: 12.68% Al=Bi: 83.75%

http://dx.doi.org/10.5772/67220

R′Al≡BiR′	WBI	NBO analysis		NRT analysis		
		Occupancy	Hybridization	Polarization	Total/covalent/ ionic	Resonance weight
		$\pi = 1.90$	π: 0.4430 Al ($sp^{59.07}$) + 0.8965 Bi ($sp^{35.38}$)	19.63% (Al) 80.37% (Bi)		Al≡Bi: 3.57%

(1) The Wiberg bond index (WBI) for the Al=Bi bond and occupancy of the corresponding σ and π bonding NBO: see Refs. [63, 64], and (2) the natural resonance theory (NRT): see Refs. [73–75].

Table 12. Selected results for the natural bond orbital (NBO) and natural resonance theory (NRT) analyses of R'Al≡BiR' compounds that have small substituents, at the B3LYP/LANL2DZ+dp level of theory [1–8, 76–80].

R′Ga≡BiR′	WBI	NBO analysis		NRT analysis		
		Occupancy	Hybridization	Polarization	Total/covalent/ ionic	Resonance weight
R′ = Tbt	2.09	$\sigma = 1.78$	σ: 0.7338 Ga ($sp^{1.04}$) + 0.6794 Bi ($sp^{31.81}$)	53.85% (Ga) 46.15% (Bi)	2.09/1.36/0.73	Ga=Bi: 11.42% Ga=Bi: 87.53%
		$\pi = 1.94$	π: 0.4586 Ga ($sp^{1.00}$) + 0.8886 Bi ($sp^{1.00}$)	21.04% (Ga) 78.96% (Bi)		Ga≡Bi: 1.05%
R′ = Ar*	2.18	$\sigma = 1.93$	σ: 0.8142 Ga ($sp^{0.12}$) + 0.5806 Bi ($sp^{40.52}$)	66.29% (Ga) 33.71% (Bi)	2.05/1.30/0.75	Ga=Bi: 10.87% Ga=Bi: 88.05%
		$\pi = 1.84$	π: 0.4810 Ga ($sp^{50.27}$) + 0.8767 Bi ($sp^{14.17}$)	23.13% (Ga) 76.87% (Bi)		Ga≡Bi: 1.08%
R′ = SiMe ($SitBu_3$)$_2$	2.26	$\sigma = 1.80$	σ: 0.7393 Ga ($sp^{0.97}$) + 0.6733 Bi ($sp^{27.39}$)	54.66% (Ga) 67.33% (Bi)	2.11/1.35/0.76	Ga=Bi: 11.84% Ga=Bi: 82.11%
		$\pi = 1.86$	π: 0.4937 Ga ($sp^{22.70}$) + 0.8696 Bi ($sp^{17.00}$)	24.38% (Ga) 75.62% (Bi)		Ga≡Bi: 6.05%
R′ = Si*i*PrDis$_2$	2.31	$\sigma = 1.90$	σ: 0.7462 Ga ($sp^{0.92}$) + 0.6657 Bi ($sp^{58.99}$)	55.68% (Ga) 44.32% (Bi)	1.93/1.40/0.53	Ga=Bi: 12.93% Ga=Bi: 80.74%
		$\pi = 1.84$	π: 0.4201 Ga ($sp^{99.99}$) + 0.9075 Bi ($sp^{99.99}$)	17.65% (Ga) 82.35% (Bi)		Ga≡Bi: 6.33%

(1) The Wiberg bond index (WBI) for the Ga=Bi bond and occupancy of the corresponding σ and π bonding NBO: see Refs. [63, 64], and (2) the natural resonance theory (NRT): see Refs. [73–75].

Table 13. Selected results for the natural bond orbital (NBO) and natural resonance theory (NRT) analyses of R'Ga≡BiR' compounds that have small substituents, at the B3LYP/LANL2DZ+dp level of theory [1–8, 76–80].

R′In≡BiR′	WBI	NBO analysis				NRT analysis
		Occupancy	Hybridization	Polarization	Total/covalent/ ionic	Resonance weight
R′ = Tbt	2.05	σ = 1.78	σ: 0.7626 In ($sp^{0.09}$) + 0.6469 Bi ($sp^{50.74}$)	58.15% (In) 41.85% (Bi)	2.10/1.15/0.95	In=Bi: 14.33% In=Bi: 76.37%
		π = 1.94	π: 0.4230 In ($sp^{99.99}$) + 0.9061 Bi ($sp^{95.23}$)	17.85% (In) 82.11% (Bi)		In≡Bi: 0.93%
R′ = Ar*	2.15	σ = 1.97	σ: 0.7145 In ($sp^{0.07}$) + 0.6996 Bi ($sp^{21.13}$)	51.05% (In) 48.95% (Bi)	2.11/1.04/1.07	In=Bi: 11.10% In=Bi: 85.07%
		π = 1.94	π: 0.4774 In ($sp^{99.99}$) + 0.8787Bi ($sp^{99.99}$)	22.79% (In) 77.21% (Bi)		In≡Bi: 3.83%
R′ = SiMe ($SitBu_3$)$_2$	2.21	σ = 1.74	σ: 0.7433 In ($sp^{0.99}$) + 0.6690 Bi ($sp^{35.62}$)	55.25% (In) 44.75% (Bi)	2.14/1.22/0.92	In=Bi: 14.20% In=Bi: 81.22%
		π = 1.88	π: 0.4468 In ($sp^{42.13}$) + 0.8696 Bi ($sp^{12.11}$)	19.96% (In) 80.04% (Bi)		In≡Bi: 4.58%
R′ = Si*i*PrDis$_2$	2.3	σ = 1.92	σ: 0.7341 In ($sp^{53.22}$) + 0.6790 Bi ($sp^{0.98}$)	46.11% (In) 53.89% (Bi)	1.88/1.27/0.61	In=Bi: 15.31% In=Bi: 81.02%
		π = 1.78	π: 0.4357 In ($sp^{26.46}$) + 0.9001 Bi ($sp^{99.99}$)	18.99% (In) 81.01% (Bi)		In≡Bi: 3.67%

Notes: (1) The Wiberg bond index (WBI) for the In≡Bi bond and occupancy of the corresponding σ and π bonding NBO: see Refs. [63, 64], and (2) the natural resonance theory (NRT): see Refs. [73–75].

Table 14. Selected results for the natural bond orbital (NBO) and natural resonance theory (NRT) analyses of R'In≡BiR' compounds that have small substituents, at the B3LYP/LANL2DZ+dp level of theory [1–8, 76–80].

R′Tl≡BiR′	WBI	NBO analysis			NRT analysis	
		Occupancy	Hybridization	Polarization	Total/covalent/ ionic	Resonance weight
R′ = Tbt	2.08	σ = 1.96	σ:0.8268 Tl ($sp^{0.07}$) + 0.5625 Bi ($sp^{99.99}$)	68.36% (Tl) 31.64% (Bi)	2.15/1.81/0.34	Tl=Bi: 18.71% Tl = Bi: 71.54%
		π = 1.94	π:0.4315 In ($sp^{99.99}$) + 0.9021 Bi ($sp^{99.99}$)	18.61% (In) 81.37% (Bi)		Tl≡Bi: 9.75%
R′ = Ar*	2.05	σ = 1.97	σ: 0.8159 Tl ($sp^{0.05}$) + 0.5782 Bi ($sp^{99.99}$)	66.57% (Tl) 33.43% (Bi)	2.11/1.70/0.41	Tl=Bi: 21.11% Tl = Bi: 70.15%
		π = 1.95	π: 0.4788 In ($sp^{99.99}$) + 0.8779 Bi ($sp^{99.99}$)	22.92% (In) 77.08% (Bi)		Tl≡Bi: 8.74%

http://dx.doi.org/10.5772/67220

R'Tl≡BiR'	WBI	NBO analysis			NRT analysis	
		Occupancy	Hybridization	Polarization	Total/covalent/ ionic	Resonance weight
R' = SiMe (Si*t*Bu3)2	2.1	σ = 1.98	σ: 0.7986 Tl ($sp^{0.02}$) + 0.6019 Bi ($sp^{31.50}$)	63.77% (Tl) 36.23% (Bi)	2.21/1.40/0.81	Tl=Bi: 16.13% Tl = Bi: 76.20%
		π = 1.92	π: 0.3890 Tl ($sp^{99.99}$) + 0.9212 Bi ($sp^{1.00}$)	15.13% (Tl) 84.87% (Bi)		Tl≡Bi: 7.67%
R' = Si*i*PrDis2	2.13	σ = 1.98	σ: 0.7757 Tl ($sp^{0.03}$) + 0.6311 Bi ($sp^{23.28}$)	60.17% (Tl) 39.83% (Bi)	1.85/1.23/0.62	Tl=Bi: 20.21% Tl=Bi: 74.69%
		π = 1.91	π: 0.4149 Tl ($sp^{99.99}$) + 0.9099 Bi ($sp^{99.99}$)	17.22% (Tl) 82.78% (Bi)		Tl≡Bi: 5.10%

Notes: (1) The Wiberg bond index (WBI) for the Tl=Bi bond and occupancy of the corresponding σ and π bonding NBO: see Refs. [63, 64], and (2) the natural resonance theory (NRT): see Refs. [73–75].

Table 15. Selected results for the natural bond orbital (NBO) and natural resonance theory (NRT) analyses of R'Tl≡BiR' compounds that have small substituents, at the B3LYP/LANL2DZ+dp level of theory [1–8, 76–80].

4. Overview of RE_{13}≡BiR (E = B, Al, Ga, In, and Tl) systems

This study of the effect of substituents on the possibilities of the existence of triply bonded RE_{13}≡BiR allows the following conclusions to be drawn (**Scheme 4**):

1. The theoretical observations strongly demonstrate that bonding mode (B) is dominant in the triply bonded RE_{13}≡BiR species, since their structures are bent to increase stability, due to electron transfer (denoted by arrows in **Figure 1**) as well as the relativistic effect [65–68].

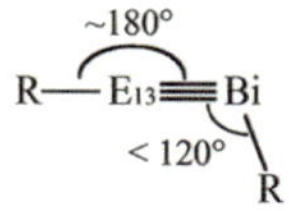

E_{13} = B, Al, Ga, In, and Tl

R = { F, OH, CH_3, H, and SiH_3 ; Tbt, Ar*, SiMe(Si*t*Bu_3)$_2$, and Si*i*PrDis$_2$ }

2. The theoretical evidence shows that both the electronic and the steric effects of substituents are crucial to making the E_{13}≡Bi triple bond synthetically accessible. Based on the present theoretical study, however, these E_{13}≡Bi triple bonds should be weak, not as strong as the traditional C≡C triple bond. From our theoretical study, both bulky and electropositive substituents, such as the silyl groups demonstrated in Scheme 1, have a significant effect on the stability of E_{13}≡Bi triply bonded compounds.

Acknowledgements

The authors are grateful to the National Center for High-Performance Computing of Taiwan in providing huge computing resources to facilitate this research. They also thank and the Ministry of Science and Technology of Taiwan for the financial support.

Author details

Jia-Syun Lu[1], Ming-Chung Yang[1], Shih-Hao Su[1], Xiang-Ting Wen[1], Jia-Zhen Xie[1] and Ming-Der Su[1,2]*

*Address all correspondence to: midesu@mail.ncyu.edu.tw

1 Department of Applied Chemistry, National Chiayi University, Chiayi, Taiwan

2 Department of Medicinal and Applied Chemistry, Kaohsiung Medical University, Kaohsiung, Taiwan, Chiayi, Taiwan

References

[1] Power P P. π-Bonding and the lone pair effect in multiple bonds between heavier main group elements. Chem. Rev. 1999; **99**:3463–3504.

[2] Power P P. Silicon, germanium, tin and lead analogues of acetylenes. Chem. Commun. 2003; **17**:2091–2101.

[3] Lein M, Krapp A, Frenking G. Why do the heavy-atom analogues of acetylene E_2H_2 (E = Si-Pb) exhibit unusual structures? J. Am. Chem. Soc. 2005; **127**:6290–6299.

[4] Sekiguchi A, Ichinohe M, Kinjo R. The chemistry of disilyne with a genuine Si–Si triple bond: synthesis, structure, and reactivity. Bull. Chem. Soc. Jpn. 2006; **79**:825–832.

[5] Power P P. Bonding and reactivity of heavier group 14 element alkyne analogues. Organometallics 2007; **26**:4362–4372.

[6] Fischer R C, Power P P. π-bonding and the lone pair effect in multiple bonds involving heavier main group elements: developments in the new millennium. Chem. Rev. 2010; **110**:3877–3923.

[7] Sasamori T, Han J S, Hironaka K, Takagi N, Nagase S, Tokitoh N. Synthesis and structure of stable 1,2-diaryldisilyne. Pure Appl. Chem. 2010; **82**:603–612.

[8] Peng Y, Fischer R C, Merrill W A, Fischer J, Pu L, Ellis B D, Fettinger J C, Herber R H, Power P P. Substituent effects in ditetrel alkyne analogues: multiple vs. single bonded isomers. Chem. Sci. 2010; **1**:461–468.

[9] Sekiguchi A, Kinjo R, Ichinohe M. A stable compound containing a silicon-silicon triple bond. Science 2004; **305**:1755–1757.

[10] Kravchenko V, Kinjo R, Sekiguchi A, Ichinohe M, West R, Balazs Y S, Schmidt A, Karni M, Apeloig Y. Solid-state ^{29}Si nmr study of RSiSiR: a tool for analyzing the nature of the Si-Si bond. J. Am. Chem. Soc. 2006; **128**:14472–14473.

[11] Sasamori T, Hironaka K, Sugiyama T, Takagi N, Nagase S, Hosoi Y, Furukawa Y, Tokitoh N. Synthesis and reactions of a stable 1,2-diaryl-1,2-dibromodisilene: a precursor for substituted disilenes and 1,2-diaryldisilyne. J. Am. Chem. Soc. 2008; **130**:13856–13857.

[12] Wiberg N, Vasisht S K, Fischer G, Mayer P. Disilynes. III [1] a relatively stable disilyne RSi≡SiR (R = SiMe(SitBu$_3$)$_2$). Z. Anorg. Allg. Chem. 2004; **630**:1823–1828.

[13] Stender M, Phillips A D, Wright R J, Power P P. Synthesis and characterization of a digermanium analogue of an alkyne. Angew. Chem. Int. Ed. 2002; **41**:1785–1787.

[14] Pu L, Phillips A D, Richards A F, Stender M, Simons R S, Olmstead M M, Power P P. Germanium and tin analogues of alkynes and their reduction products. J. Am. Chem. Soc. 2003; **125**:11626–11636.

[15] Stender M, Phillips A. D, Power P P. Formation of [Ar*Ge{$CH_2C(Me)C(Me)CH_2$}CH_2C (Me)N]$_2$ (Ar* = C_6H_3-2,6-Trip$_2$; Trip = C_6H_2-2,4,6-*i*-Pr$_3$) via reaction of Ar*GeGeAr* with 2,3-dimethyl-1,3-butadiene: evidence for the existence of a germanium analogue of an alkyne. Chem. Commun. 2002; **12**:1312–1313.

[16] Spikes G H, Power P P. Lewis base induced tuning of the Ge–Ge bond order in a "digermyne". Chem. Commun. 2007; **1**:85–87.

[17] Sugiyama Y, Sasamori T, Hosoi Y, Furukawa Y, Takagi N, Nagase S, Tokitoh N. Synthesis and properties of a new kinetically stabilized digermyne: new insights for a germanium analogue of an alkyne. J. Am. Chem. Soc. 2006; **128**:1023–1031.

[18] Phillips A D, Wright R J, Olmstead M M, Power P P. Synthesis and characterization of 2,6-Dipp$_2$-$H_3C_6SnSnC_6H_3$-2,6-Dipp$_2$ (Dipp = H_3-2,6-Pri$_2$): a tin analogue of an alkyne. J. Am. Chem. Soc. 2002; **124**:5930–5931.

[19] Pu L, Twamley B, Power P P. Synthesis and characterization of 2,6-Trip$_2H_3C_6PbPbC_6H_3$-2,6-Trip$_2$ (Trip = C_6H_2-2,4,6-*i*-Pr$_3$): a stable heavier group 14 element analogue of an alkyne. J. Am. Chem. Soc. 2000; **122**:3524–3525.

[20] Liao H-Y, Su M-D, Chu S-Y. A stable species with a formal Ge≡C triple bond — a theoretical study. Chem. Phys. Lett. 2001; **341**:122–128.

[21] Wu P-C, Su M-D. Theoretical designs for germaacetylene (RC≡GeR): a new target for synthesis. Dalton. Trans. 2011; **40**:4253–4259.

[22] Wu P-C, Su M-D. Effects of substituents on the thermodynamic and kinetic stabilities of HCGeX (X = H, CH_3, F, and Cl) isomers. A theoretical study. Inorg. Chem. 2011; **50**:6814–6814.

[23] Wu P-C, Su M-D. A new target for synthesis of triply bonded plumbacetylene (RC≡PbR): a theoretical design. Organometallics 2011; **30**:3293–3301.

[24] Silvestru C, Breunig H J, Althaus H. Structural chemistry of bismuth compounds. I. Organobismuth derivatives. Chem. Rev. 1999; **99**:3277–3328.

[25] Bishop M B, LaiHing K, Cheng P. Y, Pesehke M, Duncan M A. Growth patterns and photoionization dynamics of In/Sb and In/Bi intermetallic clusters. J. Phys. Chem. 1989; **93**:1566–1569.

[26] Itami T, Masaki T, Kuribayasi K, Sato E, Hinada M, Yamasita M, Kawasaki K. Growth of fluctuations under microgravity in liquid Bi-Ga Alloys. J. Non-Cryst. Solids. 1996; **205–207**:375–378.

[27] Mudry S, Korolyshyn A. X-ray study of the structure of liquid Bi-T1. J. Alloys Compd. 1996; **235**:120–123.

[28] Boa D, Ansara I. Thermodynamic assessment of the ternary system Bi-In-Pb. Thermochimica Acta. 1998; **314**:79–86.

[29] Khairulin R A, Stankus S V, Sorokin A L. Determination of the two-melt phase boundary and study of the binary diffusion in liquid Bi–Ga system with a miscibility gap. J. Non-Cryst. Solids. 2002; **297**:120–130.

[30] Xi Y, Zu F-Q, Li X-F, Yu J, Liu L-J, Li Q, Chen Z-H. High-temperature abnormal behavior of resistivities for Bi–In melts. Phys. Lett. 2004; **329**:221–225.

[31] Mangalam R V K, Ranjith R, Iyo A, Sundaresan A, Krupanidhi S B, Rao C N R. Ferroelectricity in $Bi_{26-x}M_xO_{40-\delta}$ (M = Al and Ga) with the γ -Bi_2O_3 structure. Solid State Commun. 2006; **140**:42–44.

[32] Brunetti B, Gozzi D, Iervolino M, Piacente V, Zanicchi G, Parodi N, Borzone G. Bismuth activity in lead-free solder Bi–In–Sn alloys. Comput Coupling Phase Diagr. Thermochem. 2006; **30**:431–442.

[33] Paliwal M, Jung I H. Thermodynamic modeling of the Al–Bi, Al Sb, Mg–Al–Bi and Mg–Al–Sb systems. Comput Coupling Phase Diagr. Thermochem. 2010; **34**:51–63.

[34] Gupta U, Reveles J U, Melko J J, Khanna S N, Castleman Jr. A. W. Origins of stability in mixed bismuth-indium clusters. J. Phys. Chem. C 2010; **114**:15963–15972.

[35] Aksöz S, Ocak Y, Maraşlı N, Keşlioğlu K. Thermal conductivity and interfacial energy of solid Bi solution in the Bi–Al–Zn eutectic system. Fluid Phase Equilib. 2010; **293**:32–41.

[36] Barbier J, Penin N, Cranswick L M. Melilite-type borates $Bi_2ZnB_2O_7$ and $CaBiGaB_2O_7$. Chem. Mater. 2005; **17**:3130–3136.

[37] Curtarolo S, Kolmogorov A N, Cocks F H. High-throughput ab initio analysis of the Bi–In, Bi–Mg, Bi–Sb, In–Mg, In–Sb, and Mg–Sb systems. Comput. Coupling Phase Diagr. Thermochem. 2005; **29**:155–161.

[38] Li Z, Knott S, Mikula A. Calorimetric investigations of liquid Bi–In–Zn alloys. Acta Materialia 2007; **55**:2417–2422.

[39] Madouri D, Ferhat M. How do electronic properties of conventional III–V semiconductors hold for the III–V boron bismuth BBi compound? Phys Stat Sol. 2005; **242**:2856–2863.

[40] Gave M A, Malliakas C D, Weliky D P, Kanatzidis M G. Wide compositional and structural diversity in the system Tl/Bi/P/Q (Q = S, Se) and observation of vicinal P–Tl J coupling in the solid state. Inorg. Chem. 2007; **46**:3632–3644.

[41] Witusiewicz V T, Hecht U, Böttger B, Rex S. Thermodynamic re-optimisation of the Bi–In–Sn system based on new experimental data. J. Alloys Compounds. 2007; **428**:115–124.

[42] Sun Z, Zhu Z, Gao Z, Tang Z. Experimental and theoretical investigation on binary anionic clusters of Al(m)Bi(n)(-). Rapid Commun. Mass Spectrom. 2009; **23**:2663–2668.

[43] Ropero-Vegaa J L, Rosas-Barrerab K L, Pedraza-Avellaa J A, Laverde-Catañob D A, Pedraza-Rosasb J E, Niño-Gómeza M E. Photophysical and photocatalytic properties of Bi_2MNbO_7 (M = Al, In, Ga, Fe) thin films prepared by dip-coating. Mater. Sci. Eng. 2010; **174**:196–199.

[44] Geng L, Cheng W D, Lin C S, Zhang W L, Zhang H, He Z Z. Syntheses, crystal and electronic structures, and characterizations of quaternary antiferromagnetic sulfides: Ba_2MFeS_5 (M = Sb, Bi). Inorg. Chem. 2011; **50**:2378–2384.

[45] Rosas-Barrera K L, Pedraza-Avella J A, Ballén-Gaitán B P, Cortés-Peña J, Pedraza-Rosas J E, Laverde-Cataño D A. Photoelectrolytic hydrogen production using Bi_2MNbO_7 (M = Al, Ga) semiconductor film electrodes prepared by dip-coating. Mater. Sci. Eng. 2011; **176**:1359–1363.

[46] Kobayashi K, Nagase S. Silicon–silicon triple bonds: do substituents make disilynes synthetically accessible? Organometallics 1997; **16**:2489–2491.

[47] Nagase S, Kobayashi K, Takagi N. Triple bonds between heavier group 14 elements. A theoretical approach. J. Organomet. Chem. 2000; **611**:264–271.

[48] Kobayashi K, Takagi N, Nagase S. Do bulky aryl groups make stable silicon–silicon triple bonds synthetically accessible? Organometallics 2001; **20**:234–236.

[49] Takagi N, Nagase S. Substituent Effects on germanium–germanium and tin–tin triple bonds. Organometallics 2001; **20**:5498–5000.

[50] Takagi N, Nagase S. A silicon–silicon triple bond surrounded by bulky terphenyl groups. Chem. Lett. 2001; **30**:966–967.

[51] Takagi N, Nagase S. Theoretical study of an isolable compound with a short silicon-silicon triple bond, $(tBu_3Si)_2MeSiSi{\equiv}SiSiMe(SitBu_3)_2$. Eur. J. Inorg. Chem. 2002;**2002**: 2775–2778.

[52] Frisch M J, Trucks G W, Schlegel H B, Scuseria G E, Robb M A, Cheeseman J R, Scalmani G, Barone V, Mennucci B, Petersson G A, Nakatsuji H, Caricato M, Li X, Hratchian H P,

Izmaylov A F, Bloino J, Zheng G, Sonnenberg J L, Hada M, Ehara M, Toyota K, Fukuda R, Hasegawa J, Ishida M, Nakajima T, Honda Y, Kitao O, Nakai H, Vreven T, Montgomery Jr. J A, Peralta J E, Ogliaro F, Bearpark M, Heyd J J, Brothers E, Kudin K N, Staroverov V N, Keith T, Kobayashi R, Normand J, Raghavachari K, Rendell A, Burant J C, Iyengar S S, Tomasi J, Cossi M, Rega N, Millam J M, Klene M, Knox J E, Cross J B, Bakken V, Adamo C, Jaramillo J, Gomperts R, Stratmann R E, Yazyev O, Austin A J,Cammi R, Pomelli C, Ochterski J W, Martin R L, Morokuma K, Zakrzewski V G, Voth G A, Salvador P, Dannenberg J J, Dapprich S, Daniels A D, Farkas O, Foresman J B, Ortiz J V, Cioslowski J, Fox D J. Gaussian, Inc., Wallingford, CT; 2013.

[53] Zhao Y, Truhlar D G. Density functionals with broad applicability in chemistry. Acc. Chem. Res. 2008; **41**:157–167.

[54] Becke A D. Density-functional exchange-energy approximation with correct asymptotic behavior. Phys. Rev. A 1988; **38**:3098–3100.

[55] Becke A D. Density-functional thermochemistry. III. The role of exact exchange. J. Chem. Phys. 1993; **98**:5648–5652.

[56] Lee C, Yang W, Parr R G. Development of the Colle-Salvetti correlation-energy formula into a functional of the electron density. Phys. Rev. B 1998; **37**:785–789.

[57] Perdew J P, Wang Y. Accurate and simple analytic representation of the electron-gas correlation energy. Phys. Rev. B 1992; **45**:13244–13249.

[58] Weigend F, Ahlrichs R. Balanced basis sets of split valence, triple zeta valence and quadruple zeta valence quality for H to Rn: design and assessment of accuracy. Phys. Chem. Chem. Phys. 2005; **7**:3297–3305.

[59] Hay P J, Wadt W R. Ab initio effective core potentials for molecular calculations. Potentials for the transition metal atoms Sc to Hg. J. Chem. Phys. 1985; **82**:270–283.

[60] Hay P J, Wadt W R. Ab initio effective core potentials for molecular calculations. Potentials for main group elements Na to Bi. J. Chem. Phys. 1985; **82**:284–298.

[61] Hay P J, Wadt W R. Ab initio effective core potentials for molecular calculations. Potentials for K to Au including the outermost core orbitals. J. Chem. Phys. 1985; **82**:299–310.

[62] Check C E, Faust T O, Bailey J M, Wright B J, Gilbert T M, Sunderlin L S. Addition of polarization and diffuse functions to the LANL2DZ basis set for p-block elements. J. Phys. Chem. A 2001; **105**:8111–8116.

[63] Wiberg K B. Application of the pople-santry-segal CNDO method to the cyclopropylcarbinyl and cyclobutyl cation and to bicyclobutane. Tetrahedron 1968; **24**:1083–1096.

[64] Reed A E, Curtiss L A, Weinhold F. Intermolecular interactions from a natural bond orbital, donor-acceptor viewpoint. Chem. Rev. 1988; **88**:899–926.

[65] Pyykkö P, Desclaux J-P. Relativity and the periodic system of elements. Acc. Chem. Res. 1979; **12**:276–281.

[66] Kutzelnigg W. Chemical bonding in higher main group elements. Angew. Chem. Int. Ed. Engl. 1984; **23**:272–295.

[67] Pyykkö P. Relativistic effects in structural chemistry. Chem. Rev. 1988; **88**:563–594.

[68] Pyykkö P. Strong closed-shell interactions in inorganic chemistry. Chem. Rev. 1997; **97**:597–636.

[69] Allen L C. Electronegativity is the average one-electron energy of the valence-shell electrons in ground-state free atoms. J. Am. Chem. Soc. 1989; **111**:9003–9014.

[70] Jones C. Recent developments in low coordination organo-antimony and bismuth chemistry. Coord. Chem. Rev. 2001; **215**:151–169.

[71] Geng L, Cheng W D, Lin C S, Zhang W L, Zhang H, He Z Z. Syntheses, crystal and electronic structures, and characterizations of quaternary antiferromagnetic sulfides: Ba_2MFeS_5 (M = Sb, Bi). Inorg. Chem. 2011; **50**:5679–5686.

[72] Chattopadhyay A, Das K K. Electronic states of TlX (X = As, Sb, Bi): a configuration interaction study. J. Phys. Chem. A 2004; **108**:7306–7317.

[73] Glendening E D, Weinhold F. Natural resonance theory: I. General formalism. J. Comp. Chem. 1998; **19**:593–609.

[74] Glendening E D, Weinhold F. Natural resonance theory: II. Natural bond order and valency. J. Comp. Chem. 1998; **19**:610–627.

[75] Glendening E D, Badenhoop J K, Weinhold F. Natural resonance theory: III. Chemical applications. J. Comp. Chem. 1998; **19**:628–646.

[76] Bino A, Ardon M, Shirman E. Formation of a carbon-carbon triple bond by coupling reactions in aqueous solution. Science 2005; **308**:234–235.

[77] Su P, Wu J, Gu J, Wu W, Shaik S, Hiberty P C. Bonding conundrums in the c2 molecule: a valence bond study. J. Chem. Theory Comput. 2011; **7**:121–130.

[78] Ploshnik E, Danovich D, Hiberty P C, Shaik S. The nature of the idealized triple bonds between principal elements and the σ origins of trans-bent geometries—a valence bond study. J. Chem. Theory Comput. 2011; **7**:955–968.

[79] Danovich D, Bino A, Shaik S. Formation of carbon–carbon triply bonded molecules from two free carbyne radicals via a conical intersection. J. Phys. Chem. Lett. 2013; **4**:58–64.

[80] Kuczkowski A, Thomas F, Schulz S, Nieger M. Synthesis and X-ray crystal structures of novel Al–Bi and Ga–Bi compounds. Organometallics 2000; **19**:5758–5762.

Chapter 5

Physicochemical Properties and Catalytic Applications of Iron Porphyrazines and Phthalocyanines

Tomasz Koczorowski, Wojciech Szczolko and
Tomasz Goslinski

Additional information is available at the end of the chapter

http://dx.doi.org/10.5772/68071

Abstract

Porphyrazines and phthalocyanines belong to porphyrinoids, which are macrocyclic compounds consisting of four pyrrole or indole rings, respectively. The aromatic rings of porphyrazines and phthalocyanines are fused together by azamethine bridges (meso nitrogen atoms) in place of methine bridges present in porphyrins. The physicochemical properties of these macrocycles can be modified in two ways. The first is by substitution of metal cation in the core, whereas the second relies on peripheral modification with various substituents. Porphyrazines and phthalocyanines can be modified inside the macrocyclic core with various transition metal cations, including iron(II/III), which impacts their electrochemical properties and influences potential applications in redox reactions. Due to their unique optical and electrochemical properties, porphyrazines and phthalocyanines found many potential and practical applications in medicine and technology. They were mainly researched as photosensitizers in photodynamic therapy, as sensors in biomedical and analytical applications or as building blocks for materials chemistry. This chapter presents physicochemical properties and catalytic applications of iron porphyrazines and phthalocyanines. The first part summarizes the influence of peripheral and axial substituents of iron(II/III) porphyrazines and phthalocyanines on their spectral properties, whereas the second focuses on the electrochemical properties of these molecules. The third part covers the activity of selected iron(II/III) porphyrazines and phthalocyanines of potential value for diverse applications including catalytic reactions.

Keywords: catalytic properties, electrochemistry, iron, porphyrazines, phthalocyanines

1. Introduction to porphyrinoids

Porphyrinoids are macrocyclic compounds consisting of four pyrrole rings usually linked together through methine or azamethine bridges. Porphyrins (Ps) are planar and aromatic with nominally 22π-electrons of which 18π-electrons are engaged in a conjugative path. Porphyrins can be substituted at the peripheral β and methine meso positions. Chlorins and bacterichlorins possess similar structure to porphyrins, and they are defined as dihydro or tetrahydro derivatives of porphyrins. Corrole macrocycle constitutes an 18π-electron system with the characteristic feature being the lack of a methine bridge between the A and D pyrrole rings. In addition, corolles carry inside the macrocyclic core three NH protons, which is different from porphyrins and chlorins, which carry two NH protons. It is worth noting that structurally related corrins with the key natural product cobalamine (vitamin B_{12}) are not aromatic and contain only one NH proton inside the macrocyclic core. Porphyrazines (Pzs) and phthalocyanines (Pcs) commonly known as tetraazaporphyrins belong to synthetic porphyrinoids. Methine bridges are replaced by azamethine (with nitrogen atoms) as the most notable feature of their structure. In addition, Pcs are tetrabenzo tetraazaporphyrins, which have annulated benzene rings in comparison with the Pz core. Formally, in phthalocyanines, unlike in porphyrazines, the azamethine bridges combine four indole instead of pyrrole rings, respectively (**Figure 1**). The derivatives of Pcs can be obtained by substitution of fused benzo-rings at peripheral (2,3 or β) and nonperipheral positions (1,4 or α). The tetraazaporphyrins possess unique physicochemical properties due to the presence of a conjugated system of

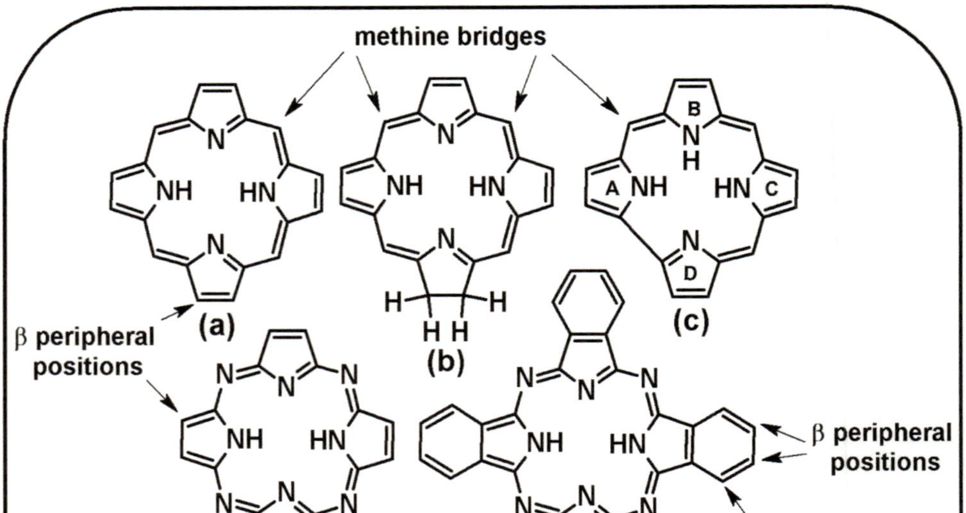

Figure 1. Structures of porphyrins (a), chlorins (b), corroles (c), porphyrazines (d), and phthalocyanines (e).

http://dx.doi.org/10.5772/68071

π electrons, bulky periphery and an ability to coordinate various metal cations inside macrocyclic core. The chelation reaction of the inner NH protons with various metal ions leads to metal chelates, which is a common feature for all porphyrinoids [1–6].

Many natural porphyrins and chlorins, for example, heme in hemoglobin and chlorophyll, reveal various vital functions and are responsible for many biochemical processes. Nowadays, porphyrinoids possess potential applications in science and technology, for example, synthetic derivatives of porphyrins as well as phthalocyanines found many applications in the dye industry and revealed potential for medicine (photodynamic therapy and photodynamic diagnosis) and technology (artificial enzymes, catalysts). Many tetraazaporphyrins have also been studied as analytical indicators, structural elements in materials chemistry, in optical data drivers and microchips, as well as photovoltaic cells. In addition, there is a possibility to utilize porphyrazines and phthalocyanines as catalysts in various organic synthesis reactions due to their ability to coordinate transition metal cations inside macrocyclic core or in the periphery. Lately, tetraazaporphyrins have also been considered as building blocks in nanotechnology due to their self-assembly and self-organization ability (**Figure 2**) [3, 6–19].

Tetraazaporphyrins can be modified using two approaches. The first relies on the introduction of various alkyl and/or aryl substituents with sulfur, nitrogen or oxygen atoms into porphyrazine β peripheral positions of pyrrole rings and phthalocyanine α nonperipheral and/or β peripheral positions of indole rings. The second concerns removal or exchange of central metal cation present in macrocyclic core. By using all of these modification approaches, there is a possibility to obtain macrocyclic compounds of altered physicochemical properties, for example, extended thermal and photochemical stability, increased solubility in organic solvents, improved luminescence and spectroscopic, magnetic, electrochemical properties, photoconductivity and surface activity [1, 2]. Incorporation of iron(II/III) cations into the porphyrinoid core allowed the application of these compounds as catalysts in redox reactions. There is a great interest in the catalytic properties of iron(II/III) tetraazaporphyrins, which dates back to the 1980s. These macrocyclic systems were considered as potential electron

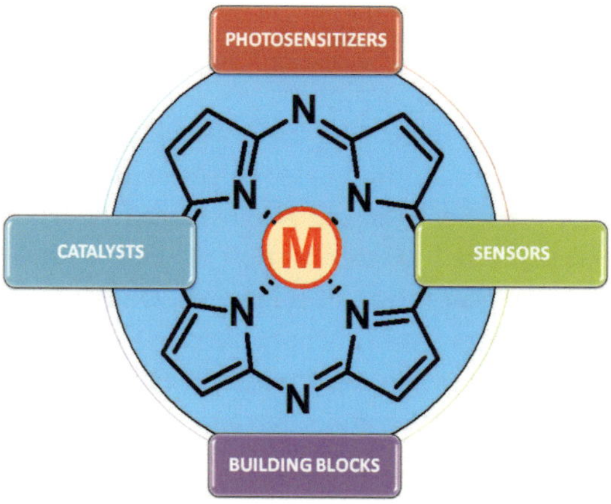

Figure 2. The main practical and potential applications of tetraazaporphyrins, M—metal ion.

and/or molecule carriers. For example, iron(III) octaphenylporphyrazine pyridine adduct developed by Stuzhin was found to be a molecular oxygen carrier [20]. Theoretical calculations using density functional theory (DFT) and experimental studies indicated that there are significant differences between metalated tetraazaporphyrins and porphyrins. The difference in the core size and shape of the macrocycle has a substantial effect on the electronic structure and properties of the overall system. DFT calculations indicated on differences in bond lengths between pyrrole/indole nitrogen atoms and coordinated iron(II) cation in porphyrins, phthalocyanines and porphyrazines, which were 1.98; 1.93 and 1.90 Å, respectively. The smaller coordination cavity results in a stronger ligand field in Pzs than in porphyrins. However, the benzo annulation in phthalocyanines produces a surprisingly strong destabilizing effect on the metal-macrocycle bonding [21, 22]. The calculations also showed how the differences in porphyrinoid (Ps, Pcs and Pzs) structures influence the axial ligand coordination of pyridine and CO to the iron(II) complexes [22].

2. Physicochemical properties of iron porphyrazines and phthalocyanines

2.1. Low solubility and tendency to form aggregates hampering utilization

Applications of porphyrazines and phthalocyanines in science and technology are limited by their low solubility in water and organic solvents and their tendency to form aggregates. These unwelcome features are the result of their likelihood to molecular interactions based on π-π stacking. Unfavorable common feature of iron(II/III) porphyrazines and phthalocyanines to form aggregates is mainly related to their conjugated, extended system of π-electrons and an ability of iron cation to coordinate compounds with heteroatoms. Annulation of a porphyrazine macrocyclic system with four benzene rings leads to a phthalocyanine with four indole rings of enhanced aggregation properties. Most of the iron(II/III) tetraazaporphyrins form two types of aggregates: *J*-type aggregates (*head to tail*) or *H*-type aggregates (*face to face*) [23, 24]. In addition, it was also observed that an incorporation of bulky substituents into iron(II) porphyrazine ring, which resulted in the separation of molecules at the distance of ca. 11 Å in the X-ray structure, did not prevent its tendency to form aggregates in solution [25].

Generally, the unsubstituted tetraazaporphyrins possess low solubility. For this reason, the most effective method applied for increasing their solubility is peripheral functionalization. Peripheral functionalization of these compounds with ester groups is able to increase their solubility in many organic solvents. For example, magnesium(II) porphyrazine with 4-hydroxybutylthio substituents was subjected to esterification reaction with 4-biphenylcarboxylic acid and further metalated with Fe^{2+} salt toward **1** (**Figure 3**). Unlike the parent demetalated porphyrazine, the metalated porphyrazines functionalized with eight ester groups were soluble in common organic solvents, such as chloroform, dichloromethane, tetrahydrofuran, acetone, and toluene and were insoluble in water and *n*-hexane [23].

Peripheral functionalization of iron(II/III) porphyrazines and phthalocyanines with halogen electron withdrawing groups (like -F or -CF_3) was found to improve their solubility in polar solvents like methanol or ethanol, ionization potential and their stability in catalytic

http://dx.doi.org/10.5772/68071

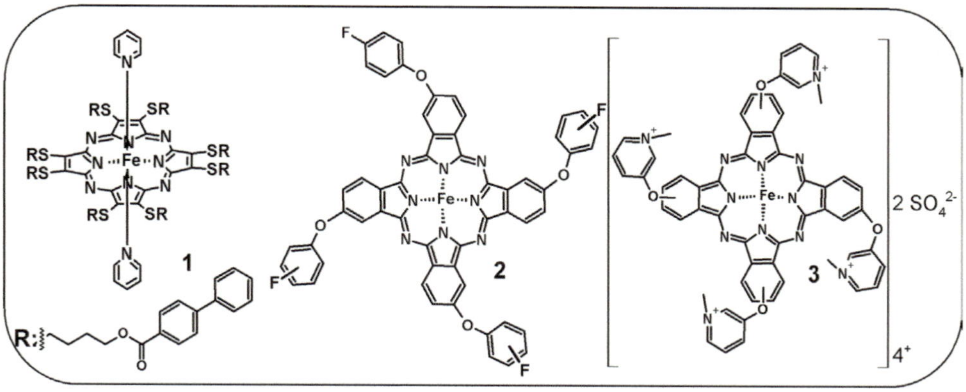

Figure 3. Structures of porphyrazine **1** and phthalocyanines **2** and **3**.

oxidation reactions [24, 26]. For example, iron(II) phthalocyanine with peripheral 4-fluorophenoxy groups **2** (**Figure 3**), which was synthesized using microwave-assisted cyclotetramerization (much faster in comparison to classical method), revealed very good solubility in various organic solvents, thus allowing solvatochromic measurements. Unfortunately, this phthalocyanine formed *H*-aggregates as compared to demetalated phthalocyanine and other metalated phthalocyanines (CuPc, Li_2Pc), especially in polar solvents [24].

Over the years, many methods have been developed in order to obtain soluble tetraazaporphyrins and to utilize them in aqueous media. This was a big challenge because as presented above these macrocycles are known for their aggregation properties. For this reason, a study was performed aiming to incorporate macrocycles into larger structures like β-cyclodextrines (β-CDs). For example, such complexes of β-CDs and iron(II) phthalocyanine **3** with peripheral quaternary pyridinium salt substituents were obtained in Kobayashi group (**Figure 3**) [27]. It is worth noting that the quaternization reaction was performed to enhance water solubility. The addition of β-CD to the aqueous solution of **3** disturbed the monomer-dimer equilibrium, as was seen by an increase in the monomer band and a decrease in the dimer band absorbance in the UV-Vis spectrum. This inclusion complex was deposited on a glassy carbon electrode. In the cyclic voltammetry studies in aqueous solution, **3** revealed very good ability toward oxygen reduction reaction (ORR) at low potential values. It indicates that the "host" compound (β-cyclodextrine) has no influence on physicochemical properties of the "guest" compound (macrocycle).

2.2. Advanced physicochemical features

Iron cation coordinated inside a macrocyclic core of porphyrazines and phthalocyanines can be involved in redox reactions and influence their electrochemical properties. By changing the valence of central iron(II/III) metal cation in tetraazaporphyrins, it is possible to transfer electrons on diverse molecules. This feature concerns also axially coordinated compounds, which form enhanced complexes and can be divided into two types. To the first group belong small ions or molecules with heteroatoms in their structure, like pyridine, pyrazine or hydroxyl and bisulfate anions. The obtained complexes are formally named as the axial complexes. To the second group belong dimers with single atom bridging groups between two iron macrocycles. In both cases, the obtained molecules have modified optical and electrochemical properties [28, 29].

A coordination of iron cation with proper ligand results in the formation of five- or six-coordinated macrocyclic complexes, which were subjected to broad study by Stuzhin et al. [28]. The coordination of iron(II) tetraazaporphyrins with axial ligands leads to the oxidation of iron(II) to iron(III). In this way, iron(II) octaphenylporphyrazine coordinated axially with F^-, Cl^-, Br^-, I^- and HSO_4^- anions was transformed to five-coordinated iron(III) complex **4** (**Figure 4**) [28, 30]. The spin state of the iron in this type of complex, which was evidenced for chloride complex, depended mainly on the nature of both the macrocyclic and axial ligands. In addition, when iron porphyrazine was coordinated with pyridine, the macrocyclic ligand adopted an intermediate structure between common porphyrins and phthalocyanines [30]. Iron(II) porphyrazines reversibly bind a variety of neutral ligands such as THF, nitrogenous bases, and carbon monoxide. This fact can be related to the higher π acidity of the porphyrazine ligand as compared to the porphyrin ligand. Iron(II) porphyrazine demonstrates no affinity for molecular oxygen, which can be a result of positively shifted III/II redox couple potential [31].

Another group of complexes, six-coordinated iron(II) complexes called bisaxial complexes also constitute a large group of compounds. DMSO, pyridine and pyrazine are one of the most often utilized molecules for coordination of iron cation, thus forming adducts as it was studied for iron(II) tetrakis(thiadiazole)porphyrazine **5** in Ercolani and Stuzhin groups (**Figure 4**) [32]. Coordinated bidentate molecules, for example, pyrazine, can be used as linkers between two or more macrocycles forming bridged complexes. What is more, the axial ligation of metalated Pz derivative with chloride influenced the Q-band absorption, which was the result of change in the symmetry from D_{4h} to C_{4v} [23]. FT-IR studies of iron(II) porphyrazine complex **5** axially coordinated with DMSO showed that two Fe-O coordination bonds

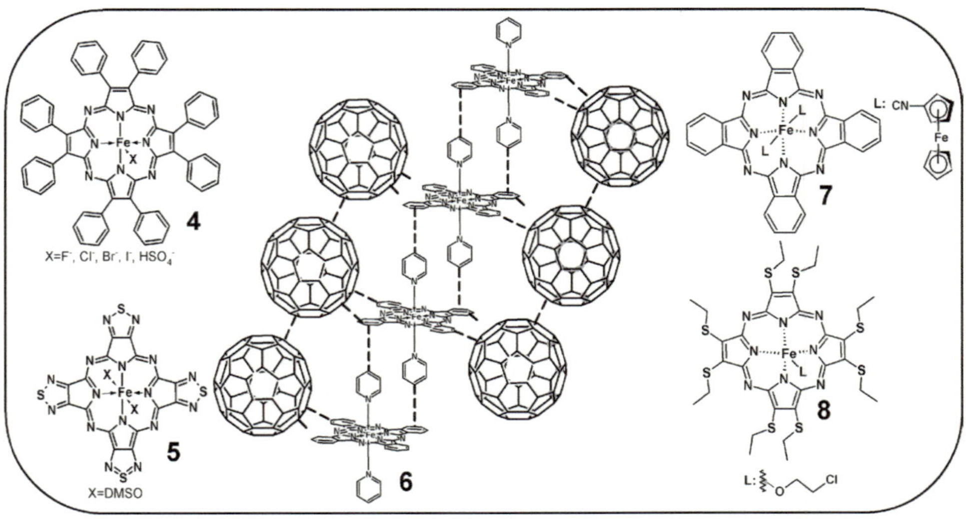

Figure 4. Chemical structures of iron(III) octaphenylporphyrazine **4**, iron(II) tetrakis(thiadiazole)porphyrazine **5**, crystal structure of **6**, phthalocyanine trinuclear molecular wire **7** and iron(III) porphyrazine **8**.

were formed. However, in phthalocyanines, analogical process was based on Fe-S coordination bond formation. It indicates that porphyrazine **5** is stronger π-acceptor [32].

Iron(II) phthalocyanines demonstrate the ability to form coordination assemblies with large structures like neutral and negatively charged fullerenes. An interesting example is crystal of **6** (**Figure 4**) obtained by cocrystallization from *n*-hexane and composed of fullerenes C_{60} and unsubstituted iron(II) phthalocyanines axially ligated with pyridines. In the structure of **6**, there are π-π stacking interactions, which do not affect the geometry of iron(II) phthalocyanine [33]. Thus, metal phthalocyanines can be involved simultaneously in molecular complexes with fullerenes as coordination assemblies and with the addition of axial ligands. Another example is trinuclear complex **7** (**Figure 4**) involving two isocyanoferrocene ligands axially coordinated to iron(II) phthalocyanine. The structure **7** forms a molecular wire between iron cation from phthalocyanine and two iron ions from ferrocenes. However, there is weak electronic communication present between two iron centers of the ferrocene ligands despite a relatively large distance (11.5 Å) [34].

An improvement in synthetic methods from the late 1980s allowed the obtaining of iron(II/III) porphyrazine and phthalocyanine complexes able to form dimers of macrocycles bridged by oxygen, nitrogen or carbon atoms (μ-oxo, μ-nitrido and μ-carbido dimers, respectively). Electrochemical studies demonstrated that macrocyclic ligand can influence the redox behavior of the binuclear complex. According to Colomban et al., dimer consisting of two porphyrazines possesses intermediate properties between corresponding porphyrin and phthalocyanine dimers [29]. This statement is based on the observation that the values of half-waved oxidation potentials of porphyrazine dimer were in the middle between similar phthalocyanine and porphyrin potentials, and for this reason, oxidation potentials in porphyrazine-based complexes are closer to phthalocyanines than porphyrins. Cyclic voltammetry of monomeric iron(III) porphyrazine axially ligated with 2-chloroethoxy substituent **8** (**Figure 4**) and its μ-oxo dimer demonstrated in DMF/TBAP three and six redox processes, respectively. Thus, the redox behavior of μ-oxoporphyrazine dimer differs significantly not only from that of the small ring system of μ-oxoporphyrin dimer, but also from that of the large-ring based μ-oxo dimer of phthalocyanine [35]. Compound **8** was also used for thin film formation utilizing Langmuir-Schaefer (LS) technique (horizontal lifting) following the study presented by Garramone et al. [36]. In aqueous solution, the μ-oxo dimer was formed as a predominant component (prevalent molecular building block) of LS films. The obtained LS-Fe films showed remarkable changes in the UV-Vis spectra, which are consistent with a significant μ-oxo dimer to monomer conversion [36]. The studies on μ-oxo iron phthalocyanine dimer demonstrated that according to the reaction conditions, it is possible to obtain bent or linear Fe-O-Fe structures (with Fe-O-Fe angle up to 300°). These two forms were identified by FT-IR signals from Fe-O antisymmetric stretching vibration. Moreover, it was possible to transform bent into linear form by applying the following reaction conditions: (i) by adding of 2-propylamine to compound suspended in chloronaphthalene or (ii) by the mixing the compound in the saturated H_2SO_4 solution in the presence of oxygen [37].

In comparison with well-known classic alone atop the other dimer structures, there is an example of significantly different "side-by-side" dimer structure **9** (**Figure 5**), which was considered as a novel molecular QCA (*quantum-dot cellular automata*) cell [38]. This dimer has

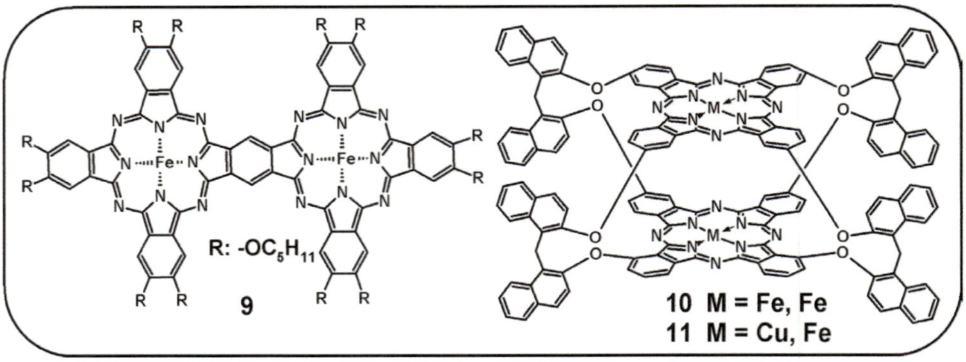

Figure 5. Structure of side-by-side dimer **9** and homonuclear and heteronulear dimers **10** and **11**.

pentyloxy substituents in the periphery, which results in its high solubility in organic solvents. The investigation of the redox properties of **9** in electrochemical study indicated that its oxidized and reduced mixed-valence complexes were stable [38]. Another of two tetraazaporphyrin macrocycles conjugated peripherally by 4,4′-[1,1′-methylenebis-(naphtalene-2,1-diyl)] bis(oxy)diphthalo-nitrile linker constitutes homonuclear (Fe-Fe) **10** and heteronuclear (Fe-Cu) **11** ball-type compounds (**Figure 5**). In both compounds, two phthalocyanine units were rigidly bound at two sides with four linking arms to form intramolecular cofacial coupling in which the splitting of the classical monophthalocyanine redox processes was observed in a cyclic voltammetry study. Moreover, the nature of the metal centers affected the distance between two Pc units in both ball-type complexes and thus the extent of the mentioned intramolecular cofacial interactions between them [39, 40].

3. Catalytic activity and electrochemical properties of iron porphyrazines and phthalocyanines

The presence of iron(II/III) cation in the coordination center of porphyrazine and phthalocyanine macrocycles determines the possibility of using them as catalysts of the oxidation-reduction reactions. Research studies carried out for several years showed that the iron tetraazaporphyrins are efficient catalysts as compared to the structurally similar porphyrin compounds. Porphyrinoid catalysts, also called biomimetic catalysts, are also more effective in carrying out the oxidation reactions of organic compounds, in comparison with other catalysts. It is related to the increased influence of the electron-donor effect of the ferric cation, which is conjugated to the π-electron system of the macrocyclic ring. The advantages of iron(II/III) porphyrazines and phthalocyanines as catalysts include high selectivity, mild and environmentally friendly reaction conditions and low energy consumption during catalysis [41]. In an early 1990s, Fitzgerald et al. provided various studies indicating significant differences in physicochemical properties of Ps, Pcs and Pzs possessing the same peripheral substituents and iron(III) cation inside a macrocyclic core [42]. It was suggested that porphyrazines are stronger σ-donors and π-acceptors than porphyrins. The electrochemical studies indicated that similarly to phthalocyanines, porphyrazines have positively shifted redox potential of 400 mV in comparison with their porphyrin analogues. Moreover, Pzs are more soluble in organic solvents than structurally relevant Pcs and can split the d orbitals of coordinated metal to a greater extent than Ps. In conclusion, it was suggested that, due to the high solubility in organic solvents, accompanied by coordination of metal ions with unusual spin states, and positively shifted redox potentials, Pzs can be considered as more efficient catalysts in comparison with Ps and Pcs [42]. Taking all this into account, iron(II/III) tetraaza-porphyrins became an object of intense studies aimed at obtaining macrostructures with increased catalytic abilities. For instance, large structures like porphyrin-phthalocyanine pentads composed of five fused macrocyclic compounds **12** were even synthesized by Kobayashi et al. (**Figure 6**) [43]. Noteworthy is that the bimetallic Fe-Cu complexes, like metal-linked face-to-face porphyrazine dimer **13** with an increased strong metal-metal spin coupling, were obtained by Barrett, Hoffman and their coworkers [44].

Iron(II/III) porphyrazines and phthalocyanines are active in redox reactions and, therefore, reveal high electrochemical activity. This feature was confirmed by cyclic voltammetry (CV) and square wave voltammetry (SWV) studies, which show, in most cases, four reversible or quasi-reversible oxidation and reduction peaks. The origin of the two peaks was attributed to reactions associated with the presence of iron cation, whereas the other two are the

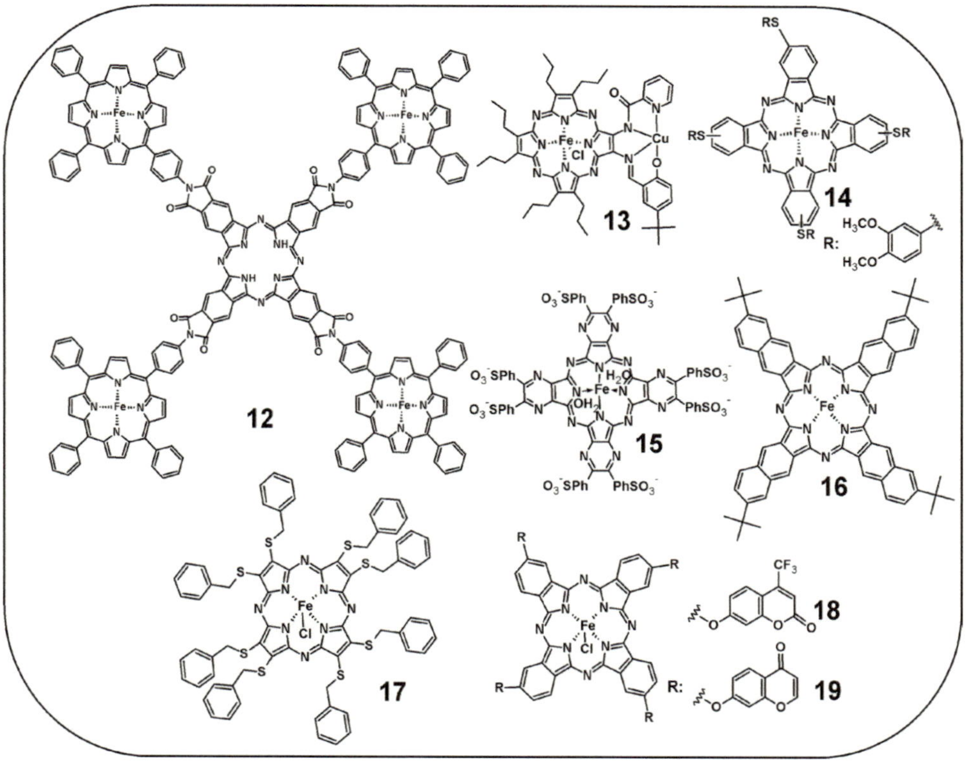

Figure 6. Structures of pentad **12**, bimetallic complex **13**, iron(II) phthalocyanine **14** and porphyrazine **15**, iron(II) naphthalocyanine **16**, iron(III) porphyrazine **17** and iron(III) phthalocyanines **18** and **19**.

result of the electronic processes within the macrocyclic ring [45]. However, there are some exceptions to this rule. For example, in the CV study performed in organic solvents for iron(II) phthalocyanine **14** (**Figure 6**), the presence of six oxidation-reduction processes was observed. Two of them were identified as reversible and irreversible reductions, while the other two were found to be quasi-reversible oxidation reactions [46]. In contrast to other iron porphyrazines, cyclic voltammetry study of water-soluble iron(II) porphyrazine **15** revealed no peaks observed for the Fe^{II}/Fe^{I} couple, probably as a result of slow heterogeneous electron transfer kinetics for this couple. Moreover, replacing of water in the coordination sphere of **15** by N-donor ligands increases the extinction coefficients of both the Q and Soret bands with a blue shift of the first one [47].

Electrochemical properties of iron(II/III) porphyrazine and phthalocyanine complexes are influenced by the periphery of the macrocycle, which can lead to an increase or a decrease of their electrochemical activity. An increase in activity is related to the presence of the peripheral substituents with lone pairs of electrons or π-electron systems, which are able to increase the coupling of electrons around the macrocycle. The decrease in activity is observed in the presence of electrochemically inactive substituents, for example, *tert*-butyl groups, as was found for iron(II) naphthalocyanine **16** (**Figure 6**) [48].

An axial coordination of molecules to the central metal ion can cause a shift of the oxidation potential of the macrocycle or a split of peaks belonging to oxidation process. The rationale for this may be connected with the coordination of solvent molecules to Fe^{3+} cation in the center of the oxidized macrocyclic compound [49]. It is known from the literature that there are differences in the values of oxidation-reduction potentials of iron phthalocyanines, when one or two solvent molecules are attached or released from the iron(II/III) cation [50].

Electrochemical studies with iron porphyrazines and phthalocyanines were also carried out using the modified electrodes with tetraazaporphyrins deposited on their surface. One example is the use of iron(III) porphyrazine **17** (**Figure 6**) as an azide and nitrate(III) anions sensor [51]. This porphyrazine was deposited onto a matrix with PVC, and the selective electrode membrane revealed a shorter electrode response time, greater tolerance within a wide range of pH at high analyte concentration and a high selectivity toward the targeted ions [51]. Another example is iron(III) phthalocyanine with four peripheral coumarin **18** or chromone **19** substituents (**Figure 6**) deposited by the electropolymerization on the Pt-working electrodes and applied as electron mediators in the electrocatalytic oxidation of nitrates(III) [52].

Studies concerning catalytic properties of iron porphyrazines and phthalocyanines have been conducted over the last 20 years, and they concerned mostly the potential applications in oxidation reactions of linear and cyclic alkenes as well as photocatalytic degradation of organic dyes. However, unsubstituted iron(III) phthalocyanine was widely used to catalyze the reaction of both the incorporation of amino substituents and the hydroxylation of aryl and alkyl molecules [53, 54]. Moreover, this compound was also used as a catalyst in the oxygen reduction reactions and revealed good stability for potential use in fuel cells or batteries [55, 56]. Some studies assessed the ability of iron(II/III) tetraazaporphyrins and their dimers in decomposition and removal of organic pollutants from industrial wastes. So far the most successfully applied photocatalytic reaction was the degradation of Rhodamine B, which was considered as a model compound in studies on environmental contamination with organic substances. The most commonly used catalysts applied were symmetrical iron sulfanylporphyrazines **20**, **21** and **22**, as well as unsymmetrical **23** (**Figure 7**). They were used either as homogeneous or heterogeneous catalysts, after deposition on the carrier, which was very often ion exchange resin (e.g., Amberlite CG400). It was shown that the deposition of the catalyst on this kind of support increases the efficiency of catalytic reactions due to an increase in the concentration of catalysts on the resin surface. Therefore, the molecules of the substrate are present in high concentration in the vicinity of the catalyst, facilitating the oxidation and reduction reactions. In addition, it was shown that the iron(II) porphyrazine—resin system is active even in the dark, without activation of macrocycle by irradiation with proper wavelength [57]. Noteworthy, some oxygen atom donors (OAD) like molecular oxygen and hydrogen peroxide were added to the reaction mixture to form an active oxygen species, which allowed the oxidative degradation of substrates. The highest reaction rate and yield were achieved with H_2O_2 [58]. It was also noticed that the modification of peripheral groups of the macrocycle by introducing hydroxymethyl substituents demonstrated an increased solubility of the catalyst in water and in organic solvents [41]. In order to improve the catalytic properties, the structure of the macrocycle was equipped with electron donor methyl moieties [59].

Another sulfur iron(II) porphyrazine **24** (**Figure 7**) was used in the catalytic oxidation reaction of organic compound, that is, X3B dye (*Reactive Brilliant Red*). The reaction was provided with hydrogen peroxide as OAD and with simultaneous exposure to light. The catalyst was active in a broad temperature and pH range, with the best yield at pH = 2, and at the higher

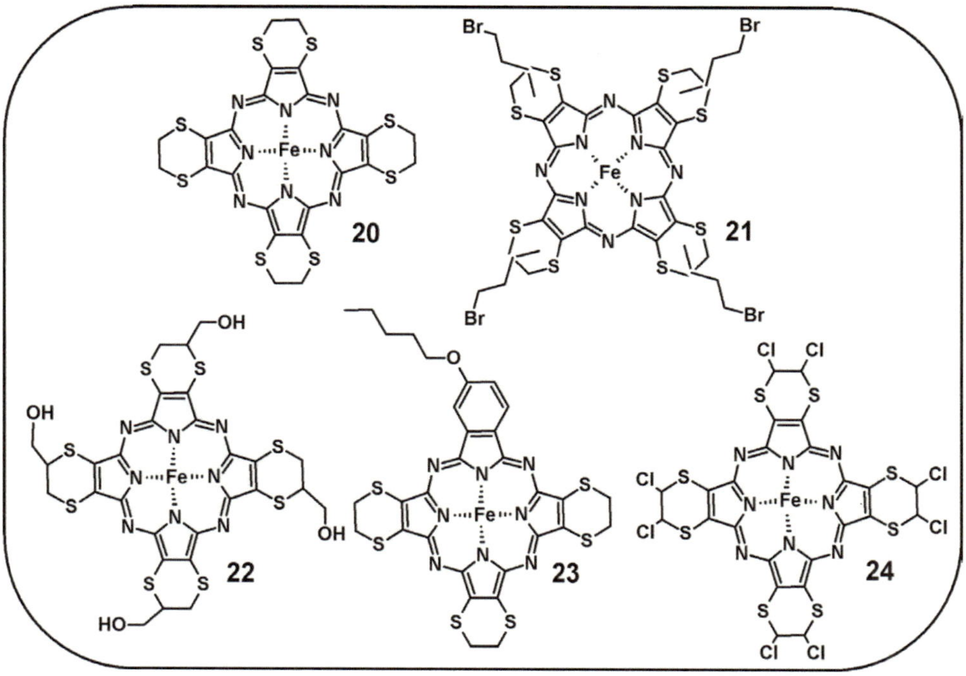

Figure 7. Chemical structure of iron(II) sulfanylporphyrazines **20–24**.

temperature. Lower catalytic ability was observed in the absence of light. It has been shown that the use of H_2O_2 as a source of oxygen resulted in the production of hydroxyl radicals as reactive oxygen species responsible for degradation of the substrate in the catalytic reaction. This process was called catalytic wet hydrogen peroxide oxidation (CWPO) and has potential for treatment of waste water [60]. As a result of research by Su et al. [61] and Theodoridis et al. [62], it was found that for the application of H_2O_2 as a commonly used oxygen atom donor in oxidation reactions catalyzed by iron(II) porphyrazine results in the hydroperoxo complex of iron(III) Pz formation. According to the reaction conditions there are two competitive redox routes: heterolysis (involving transfer of 2 electrons) and homolysis (1 electron transfer) leading to O–O bond cleavage (**Figure 8**). At acidic conditions, proton is utilized in heterolysis of O–O bond and transient high-valence iron-centered oxidizing species $Pz^{\bullet+}Fe^{IV}$=O is generated as the reactive oxygen species (ROS). Of note is that the electronic structure of N4-ligand complexes allows for the stabilization of transient high-valent intermediates. For this reason, high-valence state iron species are often identified as ROS in biomimetic catalysis. On the other hand, homolysis of O–O bond in neutral and alkaline pH conditions leads to hydroxyl radical species formation as ROS, whereas the hydroperoxo complex of porphyrazine is transformed to porphyrazine radical Fe^{III}=$O^{\bullet}$ [61].

Another important objective for the application of iron phthalocyanines and porphyrazines as catalysts of the oxidation reactions of organic compounds is their use in chemical synthesis, which leads to new derivatives without using classical synthetic routes. The presence of Fe(II/III) tetraazaporphyrins with the use of suitable oxygen donors permits one or two electron oxidation reactions. As the result, various derivatives containing epoxy groups or hydroxyl, carbonyl and carboxyl substituents can be obtained. In the study aiming to assess catalytic properties of iron porphyrazines and phthalocyanines, cyclohexane was considered as a reference compound. In various studies, there were applied iron(II) phthalocyanine derivatives **25–27** and iron(III) phthalocyanine **28** as catalysts (**Figure 9**) and also diverse sources of oxygen: *tert*-butylhydroperoxide (TBHP), hydrogen peroxide, chloro-peroxybenzoic acid (m-CPBA), molecular oxygen and oxone. It was concluded that with an increase in temperature and catalyst concentration, an increase in reaction yield was observed. The source of

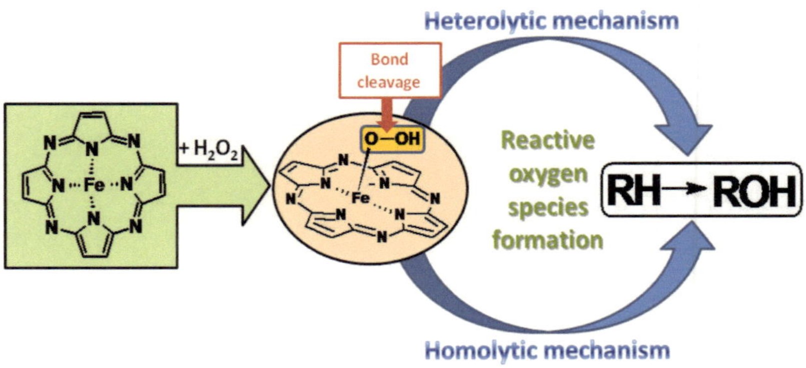

Figure 8. The heterolytic and homolytic mechanisms of O–O bond cleavage in the hydroperoxo complex of iron(III) porphyrazine following [61].

http://dx.doi.org/10.5772/68071

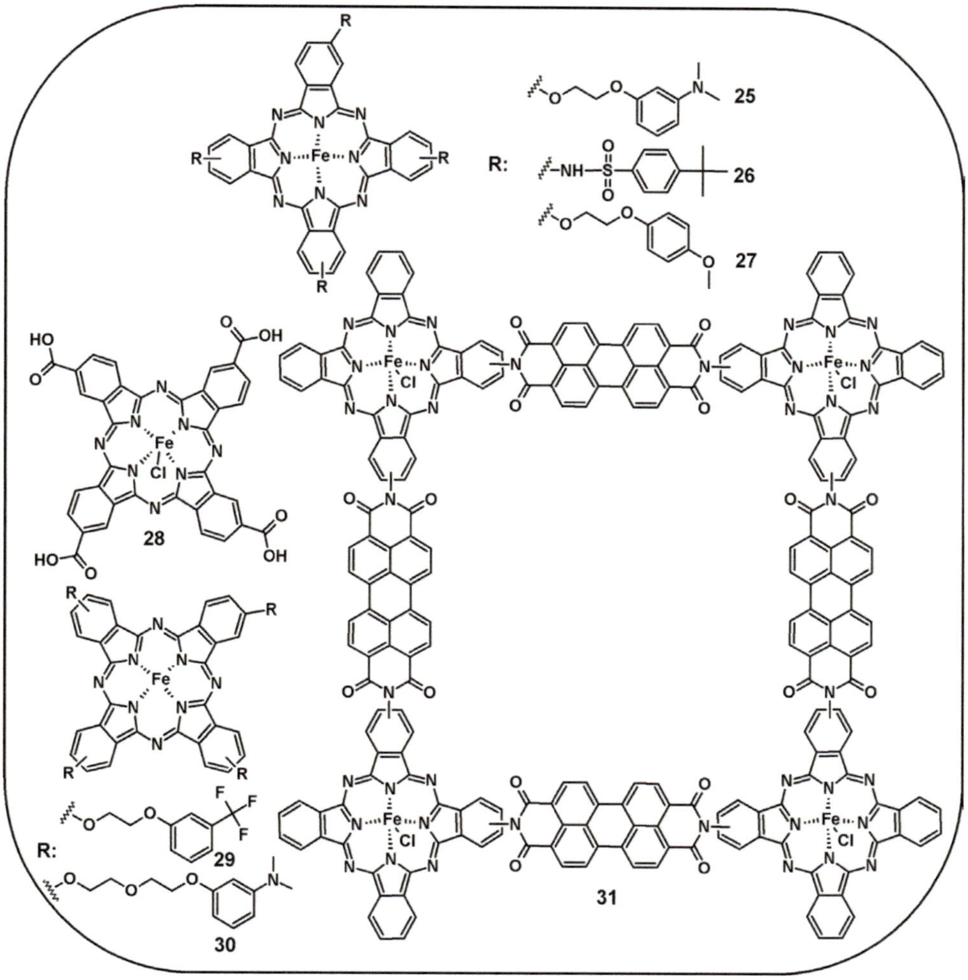

Figure 9. Structures of iron(II/III) phthalocyanine derivatives **25–30**, and cyclic tetramer **31**.

oxygen influenced the structures of obtained products. In all experiments monitored by the UV-Vis spectra, the degradation of the catalysts, manifested by a decrease of their Q-band absorption, was observed. However, in most cases, the catalytic processes proceeded, even after catalyst degradation, which can indicate the influence of other oxidation mechanisms, for example, Fenton reaction [63–66].

Lately performed study with iron(II) phthalocyanine, **29**, **30**, and the cyclic tetramer consisting of iron(III) phthalocyanine linked to 3,4,9,10-perylenetetracarboxylate, tentatively named FePPOP **31** (**Figure 9**) indicate the possibility to use these compounds toward catalytic oxidation reaction of benzyl alcohol. In the case of tetramer, it was found that **31** exhibited high stability and also exhibited the large turnover number of the reaction reaching the value of 960 [67–69]. However, in the initial phase of the experiment, the catalyst **31** acted much slower in comparison with the other compounds [69]. In addition, heterogeneous catalysts consisting of unsubstituted iron(II) phthalocyanines deposited on the electrodes by electropolymerization were applied in oxidation reactions of organic compounds. Such reactions were carried out in two kinds of systems: (i) phenolic resin/unsubstituted iron(II) phthalocyanine or (ii) the phenolic resin/structurally branched iron(II) phthalocyanine [70, 71]. In another study, various thiol derivatives (e.g., 2-mercaptoethanol) were applied as substrates in electrooxidation reaction toward disulfides. Transparent indium tin oxide electrodes were modified with iron(II) tetraaminophthalocyanine [72, 73].

4. Summary

The iron(II/III) porphyrazines and phthalocyanines have interesting electrochemical properties, which were demonstrated in many valuable studies performed during the last 30 years. Moreover, many applications of these macrocycles were presented in medicine, in biomedical and analytical fields, in materials chemistry as well as in chemical synthesis. It clarifies why catalytic abilities of iron(II/III) tetraazaporphyrins became an object of intense studies. This chapter aimed to summarize the influence of peripheral substituents of iron(II/III) porphyrazines and phthalocyanines on their spectral and electrochemical properties. Electrochemical properties of iron(II/III) porphyrazine and phthalocyanine complexes are significantly influenced by the periphery of the macrocycle, which can lead to an increase or a decrease of their electrochemical activity. Similarly, an axial coordination of molecules to the central metal ion causes a shift of the oxidation potential of the macrocycle or splits peaks belonging to oxidation processes. Selected studies on iron(II/III) porphyrazines and phthalocyanines were found not only to present their interesting physicochemical features but also further perspective applications, and thus, they were discussed in more detail. What is of immense value for further applications of these molecules in materials chemistry and nanotechnology is that some macrocycles demonstrated an ability to form coordination assemblies alone or with nanostructures, including fullerenes, and molecular wires. Especially interesting are binuclear complexes based on iron(II/III) porphyrazine and phthalocyanine bridged by oxygen, nitrogen or carbon atoms. Interesting modification of classical redox processes was observed in novel potential molecular quantum-dot cellular automata cells in which phthalocyanines were connected "side-by-side" or by forming ball-type dimers in which there were utilized sophisticated linkers binding two phthalocyanine units at two sides rigidly with four linking arms. Porphyrinoid catalysts also have the designation by biomimetic catalysts, this being because they are more effective in carrying out the oxidation reactions of organic compounds to other catalysts. It is related to the increased electron-donor effect of the ferric cation, which is conjugated to the π-electron system of the macrocyclic ring. The advantages of iron(II/III) porphyrazines and phthalocyanines as catalysts include high selectivity, mild and environmentally friendly reaction conditions and low energy consumption during catalysis. Studies of catalytic properties or iron(II/III) Pzs and Pcs concerned mostly with their potential applications in oxidation reactions of linear and cyclic alkenes as well as photocatalytic degradation of organic dyes. Some studies assessed the ability of iron(II/III) tetraazaporphyrins and their dimers in decomposition and removal of organic pollutants from industrial wastes. A huge area for further application of these macrocycles results from the electrochemical studies in which iron Pzs and Pcs were deposited on the surface of electrodes and further applied as selective anions sensors. To sum up, iron(II/III) tetraazaporphyrins appear to present many interesting perspectives for biomedical and technological applications.

Acknowledgements

Authors thank the National Science Centre, Poland, for funding (grant no. 2015/17/N/NZ7/009 43).

Author details

Tomasz Koczorowski, Wojciech Szczolko and Tomasz Goslinski*

*Address all correspondence to: tomasz.goslinski@ump.edu.pl

Department of Chemical Technology of Drugs, Faculty of Pharmacy, Poznan University of Medical Sciences, Poznan, Poland

References

[1] Michel SLJ, Hoffman BM, Baum SM, Barrett AGM. Peripherally functionalized porphyrazines: novel metallomacrocycles with broad, untapped potential. In Karlin KD ed. Prog. Inorg. Chem. J. Wiley & Sons. 2001;50:473-590. doi:10.1002/0471227110.ch8

[2] Rodríguez-Morgade MS, Stuzhin PA. The chemistry of porphyrazines: an overview. J. Porphyr. Phthalocya.2004;8:1129-1165. doi:10.1142/S1088424604000490

[3] Smith KM. Comprehensive heterocyclic chemistry. In: Katritzky AR, Rees CW (Eds.). Pergamon, Oxford. 1984;4:377-442.

[4] Aviv-Harel I, Gross Z. Coordination chemistry of corroles with focus on main group elements. Coord. Chem. Rev. 2011;255:717-736. doi:10.1016/j.ccr.2010.09.013

[5] Ishii K. Functional singlet oxygen generators based on phthalocyanines. Coord. Chem. Rev. 2012;256:1556-1568. doi:10.1016/j.ccr.2012.03.022

[6] Kryjewski M, Goslinski T, Mielcarek J. Functionality stored in the structures of cyclodextrin–porphyrinoid systems. Coord. Chem. Rev. 2015;300:101-120. doi:10.1016/j.ccr.2015.04.009

[7] Piskorz J, Goslinski T, Mielcarek J. Research in the field of azaanalogues of porphyrins and their application in pigment industry. Przem. Chem. 2014;93:1855-1858. doi:10.12916/przemchem.2014.1855

[8] Zhong C, Zhao M, Goslinski T, Stern C, Barrett AGM, Hoffman BM. Porphyrazines peripherally functionalized with hybrid ligands as molecular scaffolds for bimetallic metal-ion coordination. Inorg. Chem. 2006;45:3983-3989. doi:10.1021/ic052169p

[9] Wierzchowski M, Sobotta L, Skupin-Mrugalska P, Kruk J, Jusiak W, Michael Yee M, Konopka K, Düzgünes N, Tykarska E, Gdaniec M, Mielcarek J, Goslinski T. Phthalocyanines functionalized with 2-methyl-5-nitro-1H-imidazolylethoxy and 1,4,7-trioxanonyl moieties and the effect of metronidazole substitution on photocytotoxicity. J. Inorg. Biochem. 2013;127:62-72.doi:10.1016/j.jinorgbio.2013.06.012

[10] Yano S, Hirohara S, Obata M, Hagiya Y, Ogura S-i, Ikeda A, Kataoka H, Tanaka M, Joh T. Current states and future views in photodynamic therapy. J. Photochem. Photobiol. C Photochem. Rev. 2011;12:46-67.doi:10.1016/j.jphotochemrev.2011.06.001

[11] Sobotta L, Skupin-Mrugalska P, Mielcarek J, Goslinski T, Balzarini J. Photosensitizers mediated photodynamic inactivation against virus particles. Mini-Rev. Med. Chem. 2015;15:503-521. doi:10.2174/1389557515666150415151505

[12] Piskorz J, Skupin P, Lijewski S, Korpusinski M, Sciepura M, Konopka K, Sobiak S, Goslinski T, Mielcarek J. Synthesis, physical-chemical properties and in vitro photodynamic activity against oral cancer cells of novel porphyrazines possessing fluoroalkylthio and dietherthio substituents. J. Fluorine Chem. 2012;135:265-271. doi:10.1016/j.jfluchem.2011.12.003

[13] Kryjewski M, Smigielska A, Goslinski T. Porphyrinoid macrocycles as building blocks for nanoarrays construction. Przem. Chem. 2010;89:727-733.

[14] Shultz AM, Farha OK, Hupp JT, Nguyen ST. A catalytically active, permanently microporous MOF with metalloporphyrin struts. J. Am. Chem. Soc. 2009;131:4204-4205. doi:10.1021/ja900203f

[15] Wöhrle D, Schnurpfeil G, Makarov SG, Kazarin A, Suvorova ON. Practical applications of phthalocyanines—from dyes and pigments to materials for optical, electronic and photoelectronic devices. Macroheterocycles. 2012;5:191-202. doi:10.6060/mhc2012.120990w

[16] Jiang Z, Shao J, Yang T, Wang J, Jia L. Pharmaceutical development, composition and quantitative analysis of phthalocyanine as the photosensitizer for cancer photodynamic therapy. J. Pharm. Biomed. Anal. 2014;87:98-104. doi:10.1016/j.jpba.2013.05.014

[17] Calvete M, Yang GY, Hanack M. Porphyrins and phthalocyanines as materials for optical limiting. Synth. Met. 2004:141:231-243. doi:10.1016/S0379-6779(03)00407-7

[18] Collman JP, Boulatov R, Sunderland CJ, Fu L. Functional analogues of cytochrome c cxidase, myoglobin, and hemoglobin. Chem. Rev. 2004;104:561-588. doi:10.1021/cr0206059

[19] Kocki T, Czarczynska-Goslinska B, Kocka K, Stolarska M, Wachowska D, Lijewski S, Koczorowski T, Goslinski T. Nurses and pharmacists in interdisciplinary team of health care providers in photodynamic therapy. In: Tanaka Y. (Ed.). Photomedicine. Intech. 2017.

[20] Stuzhin PA, Latos-Grazynski L, Jezierski A. Synthesis and properties of binuclear nitride-bridged iron octaphenyltetraazaporphyrin. EPR studies of dioxygen adduct formation. Transit. Met. Chem. 1989;14:341-346.doi:10.1007/BF01032506

[21] Liao M-S, Scheiner S. Comparative study of metal-porphyrins, -porphyrazines, and -phthalocyanines. J. Comput. Chem. 2002;23:1391-403. doi:10.1002/jcc.10142

[22] Liao M, Watts JD, Huang M, Box PO, State J, Uni V. FeII in different macrocycles: electronic structures and properties. J. Phys. Chem. A. 2005;109:7988-8000.doi:10.1021/jp0581476

[23] Gonca E. Metal-free, metallo-porphyrazines, monomeric bisaxial complex [FePz(py)$_2$] and the bridged complex [FePz(pyz)$_n$ with eight (4-thiobutyl-4-biphenylcarboxylate) groups. J. Coord. Chem. 2014;67:162-175.doi:10.1080/00958972.2013.867036

[24] Jung C, Beak B, Kakimoto M. Synthesis of phthalocyanines functionalized by 4-fluorophenoxy moieties and their linear optical properties. J. Ind. Eng. Chem. 2014;20:234-237. doi:10.1016/j.jiec.2013.03.042

[25] Koczorowski T, Szczolko W, Burda K, Nowak M, Dawidowska M, Teubert A, Sobotta L, Gdaniec M, Korecki J, Mielcarek J, Tykarska E, Goslinski T. Influence of bulky pyrrolyl substitent on the physicochemical properties of porphyrazines. Dyes Pigm. 2015;112:138-144. doi:10.1016/j.dyepig.2014.06.033

[26] Sugimori T, Horike S, Handa M, Kasuga K. Preparation and some properties of perfluoroalkoxy-substituted phthalocyanine complexes of iron(III), nickel(II) and zinc(II). Inorg. Chim. Acta. 1998;278:253-255. doi:10.1016/S0020-1693(98)00104-2

[27] Tasso TT, Furuyama T, Kobayashi N. Absorption and electrochemical properties of cobalt and iron phthalocyanines and their quaternized derivatives: aggregation equilibrium and oxygen reduction electrocatalysis. Inorg. Chem. 2013;52:9206-9215. doi:10.1021/ic4002048

[28] Stuzhin PA, Mahmud H, Ulrich Z, Iron octaphenyltetraazaporphyrins: synthesis and characterization of the five-coordinate complexes of iron(III) (XFe^{III}OPTAP; X=F, Cl, Br, I, HSO_4). Inorg. Chim. Acta.1995;236:131-139.doi:10.1016/0020-1693(95)04633-K

[29] Colomban C, Kudrik EV, Tyurin DV, Albrieux F, Nefedov SE, Afanasiev P, Sorokin AB. Synthesis and characterization of μ-nitrido, μ-carbido and μ-oxo dimers of iron octapropylporphyrazine. Dalt. Trans. 2015;44:2240-2251. doi:10.1039/C4DT03207A

[30] Stuzhin PA, Mal'chugina OV, Wolowiec S, Latos-Grazynski L, Berezin BD.Formation of stable σ-aryliron(III) complexes from the reaction of chloroiron(III) octaphenyltetraazaporphyrinate with aryl Grignard reagents. Mendeleev Commun. 1998;8:143-145. doi:10.1070/MC1998v008n04ABEH000974

[31] Fitzgerald JP, Lebenson JR, Wang G, Yee GT, Noll BC, Sommer RD. Iron tetraanthracenotetraazaporphyrins: synthesis, structural characterization, ligand binding properties, and unexpected selectivity of a bis-"bowl" tetraazaporphyrin. Inorg. Chem. 2008;47:4520-4530. doi:10.1021/ic702149z

[32] Bauer EM, Cardarilli D, Ercolani C, Stuzhin PA, Russo U. Tetrakis(thiadiazole)porphyrazines. 2. Metal complexes with Mn(II), Fe(II), Co(II), Ni(II), and Zn(II). Inorg. Chem. 1999;2:6114-6120. doi:10.1021/ic990855g

[33] Konarev DV, Khasanov SS, Lyubovskaya RN. Fullerene complexes with coordination assemblies of metalloporphyrins and metal phthalocyanines. Coord. Chem. Rev. 2014;262:16-36. doi:10.1016/j.ccr.2013.10.021

[34] Nemykin VN, Purchel A, Spaeth AD, Barybin MV. Probing the electronic properties of a trinuclear molecular wire involving isocyanoferrocene and iron(II) phthalocyanine motifs. Inorg. Chem. 2013;52:11004-11012. doi:10.1021/ic4011423

[35] Pietrangeli D, Garramone G, Guascito MR, Pepe A, Rosa A, Ricciardi G. Synthesis, coordination chemistry, and physico-chemical properties of the 2-chloroethoxy-iron(III)(ethylthio)porphyrazine. J. Porphyr. Phthalocya. 2013;17:870-880. doi:10.1142/S10884246135 00685

[36] Garramone G, Pietrangeli D, Ricciardi G, Conoci S, Guascito MR, Malitesta C, Cesari D, Casilli S, Giotta L, Giancane G, Valli L. Electrochemical and spectroscopic behavior of iron(III) porphyrazines in Langmuir-Schaefer films. J. Phys. Chem. B. 2008;112:11517-11528.doi:10.1021/jp803418b

[37] Kobayashi N. Dimers, trimers and oligomers of phthalocyanines and related compounds. Coord. Chem. Rev. 2002;227:129-152.doi:10.1016/S0010-8545(02)00010-3

[38] He W, Lieberman M. The synthesis and characterization of a side-by-side iron phthalocyanine dimer. J. Porphyr. Phthalocya. 2011;15:277-292. doi:10.1142/S1088424611003264

[39] Odabas Z, Dumludağ F, Özkaya AR, Yamauchi S, Kobayashi N, Bekaroğlu Ö. Novel homo- and heterobinuclear ball-type phthalocyanines: synthesis and electrochemical, electrical, EPR and MCD spectral properties. Dalt. Trans. 2010;39:8143-8152. doi:10.1039/c0dt00398k

[40] Odabas Z, Altindal A, Özkaya AR, Salih B, Bekarolu Ö. Novel ball-type homo- and hetero-dinuclear phthalocyanines with four 1,1′-methylenedinaphthalen-2-ol bridges: synthesis and characterization, electrical and gas sensing properties and electrocatalytic performance towards oxygen reduction. Sens. Actuators, B Chem. 2010;145:355-366. doi:10.1016/j.snb.2009.12.016

[41] Chen L, Zhang Z, Wang Y, Guan Y, Deng K, Lv K, Sun J, Li Z, Li M. Photocatalytic properties and electrochemical characteristic of a novel biomimetic oxygenase enzyme photocatalyst iron(II) tetrahydroxymethyl tetra(1,4-dithiin) porphyrazine for the degradation of organic pollutants. J. Mol. Catal. A Chem. 2013;372:114-120. doi:10.1016/j.molcata.2013.02.013

[42] Fitzgerald JP, Haggerty BS, Rheingold AL, May L, Brewer GA. Iron octaethyltetraazaporphyrins—synthesis, characterization, coordination chemistry, and comparisons to related iron porphyrins and phthalocyanines. Inorg. Chem. 1992;31:2006-2013. doi:10.10 21/ic00037a007

[43] Kobayashi N, Nishiyama Y, Ohya T, Sato M. Symmetrically tetra-substituted phthalocyanines. J. Chem. Soc. Chem. Commun. 1987;390-392.doi:10.1039/C39870000390

[44] Goslinski T, Zhong C, Fuchter MJ, Stern CL, White AJP, Barrett AGM, Hoffman BM. Porphyrazines as molecular scaffolds: flexible syntheses of novel multimetallic complexes. Inorg. Chem. 2006;45:3686-3694. doi:10.1021/ic060176n

[45] Adebayo AI, Nyokong T. Synthesis, spectroscopic and electrochemical properties of manganese, nickel and iron octakis-(2-diethylaminoethanethiol)-phthalocyanine. Polyhedron. 2009;28:2831-2838. doi:10.1016/j.poly.2009.06.073

[46] Arıcı M, Arıcan D, Uğur AL, Erdoğmus A, Koca A. Electrochemical and spectroelectrochemical characterization of newly synthesized manganese, cobalt, iron and copper phthalocyanines. Electrochim. Acta. 2013;87:554-566. doi:10.1016/j.electacta.2012.09.045

[47] Kudrik EV, van Eldik R, Makarov SV. Kinetics and mechanism of water substitution in the low-spin Fe(II) complex of 4-octasulfophenylpyrazinoporphyrazine. Dalt. Trans. 2004;429-435. doi:10.1039/B311695F

[48] Baker R, Wilkinson DP, Zhang J. Facile synthesis, spectroscopy and electrochemical activity of two substituted iron phthalocyanines as oxygen reduction catalysts in an acidic environment. Electrochim. Acta.2009;54:3098-3102.doi:10.1016/j.electacta.2008.12.003

[49] Demirbas Ü, Akyüz D, Barut B, Bayrak R, Koca A, Kantekin H. Electrochemical and spectroelectrochemical properties of thiadiazole substituted metallo-phthalocyanines. Spectrochim. Acta Part A Mol. Biomol. Spectrosc. 2016;153:71-78. doi:10.1016/j.saa.2015.07.105

[50] Yu B, Lever BP, Swaddle TW. Electrochemistry of metal phthalocyanines in organic solvents at variable pressure. Inorg. Chem. 2004;43:4496-4504. doi:10.1021/ic040032d

[51] Prasad R, Gupta VK, Kumar A. Metallo-tetraazaporphyrin based anion sensors: regulation of sensor characteristics through central metal ion coordination. Anal. Chim. Acta.2004;508:61-70. doi:10.1016/j.aca.2003.11.056

[52] Chohan S, Booysen IN, Mambanda A, Akerman MP. Synthesis, characterization and electrocatalytic behavior of cobalt and iron phthalocyanines bearing chromone or coumarin substituents. J. Coord. Chem. 2015;68:1829-1846.doi:10.1080/00958972.2015.1023196

[53] Paradine SM, White MC, Iron-catalyzed intramolecular allylic C–H amination. J. Am. Chem. Soc. 2012;134:2036-2039.doi:10.1021/ja211600g

[54] Bolm C, Legros J, Le Paih J, Zani L. Iron-catalyzed reactions in organic synthesis. Chem. Rev. 2004;104:6217-6254. doi:10.1021/cr040664h

[55] Li W, Wu J, Higgins D, Choi J, Chen Z. Determination of iron active sites in pyrolyzed iron-based catalysts for the oxygen reduction reaction. ACS Catal. 2012;2:2761-2768. doi:10.1021/cs300579b

[56] Johannson M, Jorgensen KA, Iron-catalyzed allylic amination, J. Org. Chem. 1994;59:214-216. doi:10.1021/jo00080a034

[57] Lei Y, Sun J, Yang C, Deng K, Wang D. Preparation and properties of sulfur-containing tetraazaporphyrin iron supported on anion-exchange resin. J. Porphyrins Phthtalocya. 2005;9:537-543.doi:10.1142/S1088424605000654

[58] Tang J, Chen L, Sun J, Lv K, Deng K. Synthesis and properties of iron(II) tetra(1,4-dithiin) porphyrazine bearing peripheral long-chain alkyl group of active end-bromine. Inorg. Chem. Commun. 2010;13:236-239. doi:10.1016/j.inoche.2009.11.020

[59] Yang C, Sun J, Deng K, Wang D. Synthesis and photocatalytic properties of iron(II)tetramethyl-tetra(1,4-dithiin)porphyrazine. Catal. Commun. 2008;9:321-326. doi:10.1016/j.catcom.2007.07.020

[60] Zhang Z, Peng Q, Sun J, Fang L, Deng K. Enhancement of catalytic activities of a biomimetic catalyst $FePz(dtnCl_2)_4$ for the wet oxidation of Brilliant Red X3B through the synergetic effect of heat and light irradiation. Ind. Eng. Chem. Res. 2013;52:13342-13349. doi:10.1021/ie4016842

[61] Su R, Sun J, Sun Y, Deng K, Cha D, Wang D. Oxidative degradation of dye pollutants over a broad pH range using hydrogen peroxide catalyzed by $FePz(dtnCl_2)_4$. Chemosphere.2009;77:1146-51.doi:10.1016/j.chemosphere.2009.08.005

[62] Theodoridis A, Maigut J, Puchta R, Kudrik EV, van Eldik R. Novel iron(III) porphyrazine complex. Complex speciation and reactions with NO and H_2O_2. Inorg. Chem. 2008;47:2994-3013. doi:10.1021/ic702041g

[63] Bıyıklıoğlu Z, Saka ET, Gökçe S, Kantekin H. Synthesis, characterization and investigation of homogeneous oxidation activities of peripherally tetra-substituted Co(II) and Fe(II) phthalocyanines: oxidation of cyclohexene. J. Mol. Catal. A Chem. 2013;378:156-163. doi:10.1016/j.molcata.2013.06.009

[64] İsci Ü, Caner C, Zorlu Y, Gürek AG, Dumoulin F, Ahsen V. Sulfonamide-substituted iron phthalocyanine: design, solubility range, stability and oxidation of olefins. Dalt. Trans. 2014;43:17916-17919. doi:10.1039/C4DT02412E

[65] Saka ET, Bıyıklıoğlu Z. Co(II) and Fe(II) phthalocyanines: synthesis, characterization and catalytic activity on cyclohexene oxidation with different oxygen source. J. Organomet. Chem. 2013;745-746:50-56. doi:10.1016/j.jorganchem.2013.07.018

[66] Sun X, Wang L, Tan Z. Improved synthesis of soluble metal-free/metal phthalocyanine tetracarboxylic acids and their application in the catalytic epoxidation of cyclohexene. Catal. Lett. 2015;145:1094-1102.doi:10.1007/s10562-015-1500-0

[67] Aktas A, Acar İ, Bıyıklıoğlu Z, Saka ET, Kantekin H. Synthesis, electrochemistry of metal-free, copper, titanium phthalocyanines and investigation of catalytic activity of cobalt, iron phthalocyanines on benzyl alcohol oxidation bearing 4-{2-[3-trifluoromethyl)phenoxy]ethoxy} groups. Synth. Met. 2014;198:212-220. doi:10.1016/j.synthmet.2014.10.022

[68] Çakır V, Saka ET, Bıyıklıoğlu Z, Kantekin H. Highly selective oxidation of benzyl alcohol catalyzed by new peripherally tetra-substituted Fe(II) and Co(II) phthalocyanines. Synth. Met. 2014;197:233-239. doi:10.1016/j.synthmet.2014.09.022

[69] Ma P, Lv L, Zhang M, Yuan Q, Cao J, Zhu C. Synthesis of catalytically active porous organic polymer from iron phthalocyanine and diimide building blocks. J. Porous Mater. 2015;22:1567-1571. doi:10.1007/s10934-015-0039-1

[70] Hiraike Y, Saito M, Niwa H, Kobayashi M, Harada Y, Oshima M, Kim J, Nabae Y, Kakimoto M. Active site formation mechanism of carbon-based oxygen reduction catalysts derived from a hyperbranched iron phthalocyanine polymer. Nanoscale Res. Lett. 2015;10:179-190. doi:10.1186/s11671-015-0881-8

[71] Isaacs M, Armijo F, Ramírez G, Trollund E, Biaggio SR, Costamagna J, Aguirre MJ. Electrochemical reduction of CO_2 mediated by poly-M-aminophthalocyanines (M=Co, Ni, Fe): poly-Co-tetraaminophthalocyanine, a selective catalyst. J. Mol. Catal. A Chem. 2005;229:249-257. doi:10.1016/j.molcata.2004.11.026

[72] Zagal JH, Gulppi M, Isaacs M, Cárdenas-Jirón G, Jesús AguirreM. Linear versus volcano correlations between electrocatalytic activity and redox and electronic properties of metallophthalocyanines. Electrochim. Acta.1998;44:1349-1357.doi:10.1016/S0013-4686(98)00257-6

[73] Zagal JH, Griveau S, Silva JF, Nyokong T, Bedioui F. Metallophthalocyanine-based molecular materials as catalysts for electrochemical reactions. Coord. Chem. Rev. 2010;254: 2755-2791. doi:10.1016/j.ccr.2010.05.001

Chapter 6

Voltammetric Analysis of Platinum Group Metals Using a Bismuth-Silver Bimetallic Nanoparticles Sensor

Charlton van der Horst, Bongiwe Silwana,
Emmanuel Iwuoha and Vernon S. Somerset

Additional information is available at the end of the chapter

http://dx.doi.org/10.5772/68132

Abstract

This study dealt with the development of a bismuth-silver bimetallic nanosensor for differential pulse adsorptive stripping voltammetry of platinum group metals (PGMs) in environmental samples. The nanosensor was fabricated by drop coating a thin bismuth-silver bimetallic film onto the active area of the screen-printed carbon electrodes. Optimization parameters such as pH, dimethylglyoxime (DMG) concentration, deposition potential and deposition time, stability test and interferences were also studied. In 0.2 M acetate buffer (pH = 4.7) solution and DMG as the chelating agent, the reduction signal for PGMs ranged from 0.2 to 1.0 ng L^{-1}. In the study of possible interferences, the results have shown that Ni(II), Co(II), Fe(III), Na^+, SO_4^{2-}, and PO_4^{3-} do not interfere with Pd(II), Pt(II), and Rh(III) in the presence of DMG with sodium acetate buffer as the supporting electrolyte solution. The limit of detection for Pd(II), Pt(II), and Rh(III) was found to be 0.07, 0.06 and 0.2 ng L^{-1}, respectively. Good precision for the sensor application was obtained with a reproducibility of 7.58% for Pd(II), 6.31% for Pt(II), and 5.37% for Rh(III) (n = 10).

Keywords: adsorptive stripping voltammetry, bismuth-silver bimetallic, platinum group metals, dimethylglyoxime, screen-printed electrode

1. Introduction

Nowadays, the pollution of surface waters with chemical contaminants is one of the most crucial environmental problems. These chemical contaminants enter rivers and streams resulting in tremendous amount of destruction to the aquatic ecosystem [1]. Heavy and platinum group metal (PGM) contaminations at trace levels in water resources present a major current environmental threat, so the detection and monitoring of these metal contaminants result in an ever-increasing demand [2, 3]. According to El Mhammedi et al. [4], heavy metal analysis has been carried out in laboratories with time-consuming sampling, transportation and storage steps. Spectroscopy instrumentation was used for the simultaneous analysis of metal ions in water and, sediment biota samples. For the detection of trace metals, electrochemical analysis has always been recognized as a powerful tool [5–7].

In electroanalysis, much attention has been dedicated to the development of mercury-free sensors the last decade [8, 9]. Various materials have been used as working electrodes for heavy metal analysis [10, 11]. One of these working electrodes is bismuth film electrodes (BiFEs) and has been widely explored as a replacement for the toxic hanging drop mercury electrodes and films [12]. In differential pulse adsorptive stripping voltammetry (DPAdSV), nanoparticles play also a very important role. Due to their unique electronic, chemical, mechanical and thermal properties of nanoparticles make them extremely attractive for heavy metal electrochemical sensors in comparison with conventional materials [13]. In comparison with traditional macro-electrodes, nanoparticle electrodes show a decreased influence of the solution resistance, an increased mass-transport rate, and a higher signal-to-noise ratio [14].

Rahman et al. [15] use gold-silver bimetallic nanoparticles for the sensing of an environmental toxin, pyrene. These bimetallic nanoparticles were prepared by chemical reduction method involving the reduction of $AgNO_3$ and $HAuCl_4$ in aqueous solution of 2% hydrazine as reducing agent, cetryl trimethyl ammonium bromide as the capping agent, deionized water as the solvent and iso-octane as a co-precipitator. In another study, Mailu et al. [16] used overoxidized-polypyrrole/Ag-Au bimetallic nanoparticles for the determination of anthracene, using square wave voltammetry as technique. Bimetallic nanoparticles have been mainly used for the construction of biosensors and electrochemical sensors for determination of PGMs [17], arsenic [18], glucose and ascorbic acid [19], nitrite [20], carcinogenic organic compounds [21] and antidepressant drug [22].

This study focuses on the development of a new working electrode, such as bismuth-silver bimetallic nanoparticles for heavy metal analysis. The work done by Van der Horst et al. [23] describes the development of a DPAdSV procedure for the determination of platinum in environmental samples. In this procedure, a glassy carbon electrode modified with a bismuth film (GCE/Bi-AgF) was constructed. The optimization of several stripping voltammetric parameters such as dimethylglyoxime (DMG) concentration, composition of supporting electrolyte, pH, deposition potential and deposition time was performed. In this study, the results obtained showed low detection limits of 0.2 ng L^{-1} for Pt(II). These bismuth-silver bimetallic nanoparticles have been utilized for the determination of Pd(II), Pt(III), and Rh(III) [17] and Pt-Rh and Pd-Rh complexes [24] and the improved detection of ascorbic acid [25].

The main aim of the work described in this chapter was to investigate the construction and application of a disposable screen-printed carbon electrode (SPCE) that utilizes bismuth-silver bimetallic nanomaterials for PGMs determination. The electrochemical behavior of PGMs in road dust and roadside soil samples was investigated by voltammetry, and the optimum instrumental conditions were defined by DPAdSV measurements. The bismuth-silver bimetallic screen-printed nanosensor was further successfully applied for the analysis of PGMs in environmental samples.

2. Materials and methods

2.1. Reagents

Sodium acetate (NaOAc), ammonia (NH_3) (25%), ammonium chloride (NH_4Cl), hydrochloric acid and nitric acid were supplied by Merck (South Africa). All precious and heavy metal standards (1000 mg L^{-1} AAS), dimethylglyoxime (DMG) were purchased from Sigma-Aldrich (South Africa). Glacial acetic acid (95%), ethanol (95%), hexamethylenetetramine (HMTA), dichloromethane, hydrazine sulfate, and formaldehyde solution were supplied by Kimix (South Africa). Poly(vinyl) alcohol (PVA) was also obtained from Sigma-Aldrich (South Africa). Next, 0.5–20 μg L^{-1} solutions of platinum group elements were prepared by diluting the corresponding standard stock solutions. A 0.01 M ammonia buffer solution (pH = 9.0) was prepared by mixing ammonium chloride with concentration ammonia and served as the supporting electrolyte. A 0.2 M sodium acetate buffer (pH = 4.7) was prepared by mixing sodium acetate with acetic acid and deionized water. The 0.01 M DMG solution was prepared in 95% ethanol and served as the chelating agent. All solutions were prepared by Milli-Q (Millipore 18 M Ohm cm) water.

2.2. Instrumentation

Differential pulse adsorptive stripping voltammetric measurements were performed using PalmSens® portable potentiostat/galvanostat, with the PS Trace program and accessories (PalmSens® Instruments BV, 3992 BZ Houten, The Netherlands). The portable potentiostat was connected to a microcomputer controlled by PS 2.1 software for data acquisition and experimental control. All the DPAdSV measurements were performed in a conventional electrochemical cell of 20.0 mL, employing the screen-printed carbon electrode modified with bismuth-silver nanoparticles (SCPE/Bi-Ag) with 4 mm diameter provided by Dropsens (Oviedo, Spain) as working electrodes [26–28]. All experiments were performed at ambient temperatures [29].

2.3. Preparation of the bismuth-silver bimetallic film

The Bi-Ag bimetallic nanoparticles were dispersed through ultrasonic vibration in a solution of *N,N*-dimethylformamide (DMF) to form a suspension. A defined quantity of the suspension was applied to a clean surface of SPCE and dried at room temperature to get a thin film on the SPCE surface [30–32]. After each voltammetric cycle, the cleaning of the Bi-Ag bimetallic nanofilm was performed by holding the potential of the electrode at +1.0 V. Traces of the remaining DMG complexes on the electrode surface were reduced and quickly desorbed at this potential. A short cleaning period of 30 s was required to refresh the electrode surface completely [33].

2.4. Procedure for the determination of PGMs

A 10 mL of 0.2 M acetate buffer (pH = 4.7) solution containing 1×10^{-5} M DMG was used as electrolyte in the cyclic and stripping voltammetric procedures. The SPCE/Bi-Ag nanosensor was immersed into the solution and an accumulation potential of −0.7 V (vs. Ag/AgCl) for Pd(II) and −0.6 V (vs. Ag/AgCl) for Pt(II), and −0.7 V (vs. Ag/AgCl) for Rh(III) was applied, while the solution was stirred. A 30 s quiet time was used, and the voltammogram was scanned from +0.8 to −1.4 V (vs. Ag/AgCl) at a scan rate of 60 mV s^{-1} for cyclic voltammetry measurements, while scanning was performed from −0.8 to −0.1 V (vs. Ag/AgCl) for adsorptive differential pulse stripping voltammetry measurements.

For dust or soil extracted solution, 1 mL aliquot of both extracted solutions was added to 9 mL of 0.2 M sodium acetate buffer (pH = 4.7) solution, containing 1×10^{-5} M DMG and 0.5 μg L^{-1} PGM standard, respectively, to give a final volume of 10 mL. The determination of Pd(II), Pt(II), and Rh(III) was performed using both adsorptive differential pulse stripping voltammetry (AdDPSV) [34]. The PGMs were introduced into the solution after the background voltammogram was recorded. All the experiments were performed in the presence of oxygen and at room temperature [35, 36].

3. Results and discussion

3.1. Electrochemical behaviors of Bi-Ag bimetallic modified electrode

The preliminary investigation of the electroactivity of the bismuth-silver bimetallic nanofilm electrode (Bi-AgFE) was done by using cyclic voltammetry (CV) and differential pulse adsorptive stripping voltammetry (DPAdSV) measurements. To obtain optimal conditions, it is very important to study the influence of supporting electrolyte, dimethylglyoxime concentration, deposition potential, deposition time, and stability test in DPAdSV mode. In this study, different electrolytes such as 0.1 M hydrochloric acid, 0.2 M sodium acetate (pH = 4.7), 0.1 M phosphate (pH = 7.0), and 0.1 M phosphate (pH = 9.0) buffers were tested as supporting electrolytes using the bismuth-silver bimetallic nanofilm electrode (Bi-AgFE). The cyclic voltammograms (CVs) of the resulting electrode obtained in the four different buffer solutions (not shown) showed that the redox response peak height was improved in the presence of 0.2 M sodium acetate buffer solution. Thus, for voltammetric measurements, a solution of acetic acid and sodium acetate was used as the optimal buffer solutions. The results obtained showed anodic peaks at −0.2 and −0.6 V (vs. Ag/AgCl) and cathodic peaks at +0.1 and +0.4 V (vs. Ag/AgCl). On closer inspection the CV results for Bi-AgFE sensor at a scan rate of 50 mV s^{-1}, it is seen that redox couples for Bi^{3+}/Bi^{2+} and Ag^{+}/Ag are present.

3.2. Effect of reagent concentration

In differential pulse adsorptive stripping voltammetric (DPAdSV) analysis, the ligand concentration in solution has a profound effect on the voltammetric peak height. Palladium has a definite adsorption voltammetric peak in acidic medium if dimethylglyoxime (DMG) is

http://dx.doi.org/10.5772/68132

used as complexing agent. Dimethylglyoxime is suggested by Georgieva and Pihlar [37] as the complexing agent if sodium acetate is used as supporting electrolyte. In this investigation, the effect of DMG concentrations on the PGMs (Pd, Pt, Rh) peak currents was examined in the range from 5×10^{-6} to 5×10^{-5} M (**Figure 1**). The effect of DMG concentration on Pd(II), Pt(II), and Rh(III) peak currents in 0.2 M sodium acetate buffer (pH = 4.7) solution has shown that concentration of 1×10^{-5} M DMG gave the best results for the SPCE/Bi-AgF sensor, and it was decided to conduct all other stripping experiments using this DMG concentration. **Figure 1** presents the results for the current responses of PGMs complexes ((e.g., $Pd(HDMG)_2$, $Pt(HDMG)_2$, $Rh(HDMG)_3$) and the evaluated potentials in different concentrations of DMG.

3.3. Deposition potential and time studies

In electroanalytical chemistry, differential pulse voltammetry (DPV) is used as an effective and common technique when the content of analyte is very low due to its sensitivity [38]. The influence of deposition potential (E_d) and time (t_d) is always important factors on the

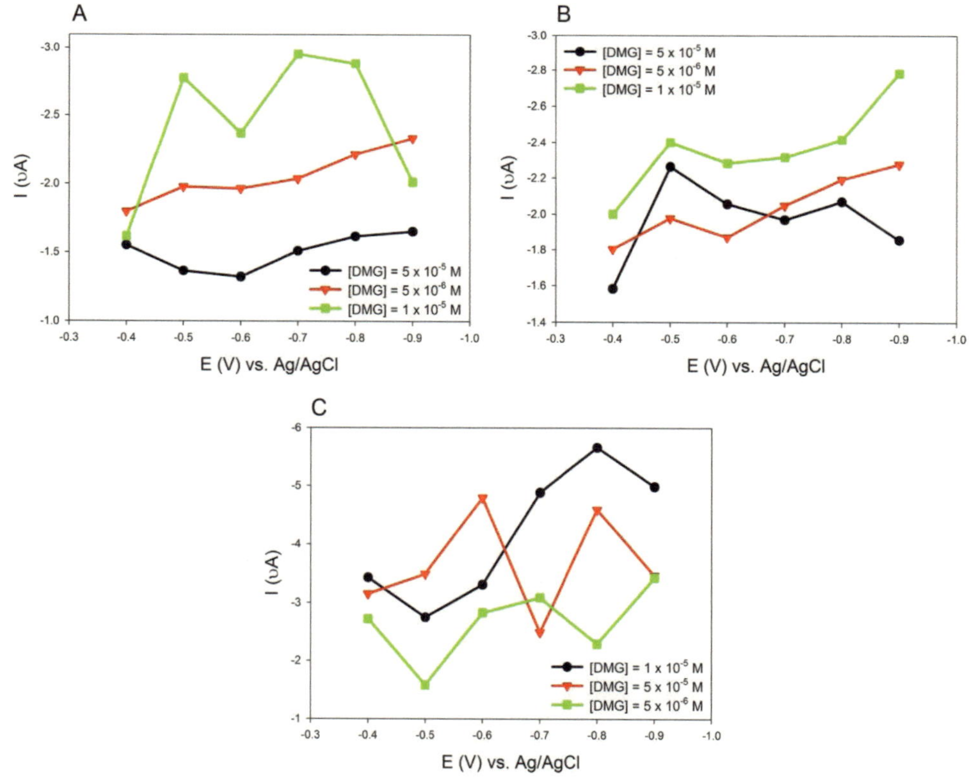

Figure 1. Effect of varying dimethylglyoxime (DMG) concentrations on the peak current results for PGMs at a SPCE/Bi-AgFE sensor. The solutions consisted of 0.2 M acetate buffer, (pH = 4.7) containing: (A) 1 ng L^{-1} Pd(II) with deposition time of 30 s V; (B) 1 ng L^{-1} Pt(II) with deposition time of 30 s; and (C) 1 ng L^{-1} Rh(III) with deposition time of 30 s. Three different concentrations of DMG were employed as demonstrated in the graphs.

sensitivity and detection limit in DPV methods. To enhance the electroanalytical performance of the Bi-Ag bimetallic sensor, the deposition potential and time were optimized. The dependence of deposition potential on the variation of stripping peak current for 1 ng L^{-1} Pd(II), Pt(II), and Rh(III) at the bismuth-silver nanosensor surface (**Figure 2A**). The optimization of deposition potential was done by varying the potential from −0.4 to +1.0 V (vs. Ag/AgCl). In the optimization results for deposition potentials have shown that for Pd(II) and Pt(II), a steady increase in the peak current responses was observed up to a E_d value of −0.7 and −0.9 V (vs. Ag/AgCl). In the case for Rh(III), a sharp increase in peak current was observed at −0.9 V (vs. Ag/AgCl). Optimum deposition potential for Pd(II), Pt(II) and Rh(III) determination in 0.2 M acetate buffer (pH = 4.7) solution is in the range from −0.7, −0.9 and −0.8 V (vs. Ag/AgCl), respectively.

The dependence of deposition time on the stripping peak current of Pd(II), Pt(II) and Rh(III) was investigated using the bismuth-silver bimetallic nanosensor (**Figure 2B**). In adsorptive stripping voltammetry (ASV), complexing agent in the electrolyte solution, after reaction, forms complex ions in the solution, and the complex is accumulated onto the sensor surface in the amount proportional to the deposition time [39]. The dependence of deposition time on the stripping peak current for Pd(II) and Rh(III) decreases almost linearly with longer deposition times. A deposition time at 30 s was chosen as the optimum deposition time in this investigation. In the case for Pt(II), stripping peak current increases with the increasing in the deposition time between 30 and 90 s and became nearly constant above 90 s due to the surface saturation of the bismuth-silver bimetallic nanosensor. In this study for all subsequent Pt(II) measurements, deposition time of 90 s was employed due to surface saturation of the bimetallic sensor.

Table 1 illustrates a summary of the optimized working conditions for the adsorptive differential pulse stripping voltammetric (AdDPSV) determination of a series of standard (or model) solutions of Pd(II), Pt(II), and Rh(III) metal ions. It was observed that Pd(II) and Rh(III) have the same deposition time with different deposition potentials. Other optimized working conditions such as DMG concentration, supporting electrolyte and potential window was the same for of Pd(II), Pt(II), and Rh(III) throughout the study.

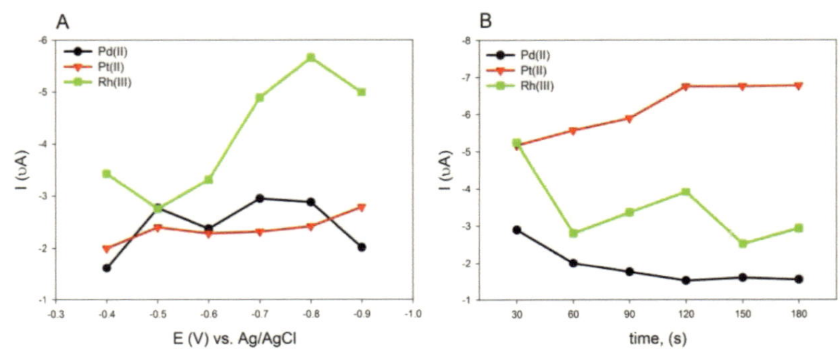

Figure 2. Results obtained for the effect of: (A) varying deposition potential (E_d) upon adsorptive stripping voltammetric responses for 1 ng L^{-1} Pd(II), Pt(II) and Rh(III); and (B) varying deposition times (t_d) upon adsorptive stripping voltammetric responses for 1 ng L^{-1} Pd(II), Pt(II) and Rh(III) at the SPCE/Bi-AgFE sensor. The solutions used consisted of 0.2 M acetate buffer (pH = 4.7) containing 1×10^{-5} M DMG concentration.

Stripping step	Determinant	$Pd(HDMG)_2$	$Pt(HDMG)_2$	$Rh(HDMG)_3$
	pH	4.7		
	Reduction potential (V)	−0.7	−0.9	−0.8
	Deposition time (s)	30	90	30
	Supporting electrolyte	0.2 M NaOAc		
Measurement	Measurement technique	DPSV		
	Potential window	0.1 to −0.6 V		
	Supporting electrolyte	0.2 M NaOAc		
	DMG concentrations (M)	Current (μA)		
	5×10^{-6}	−2.038	−2.278	−4.600
	1×10^{-5}	−2.950	−2.784	−5.660
	5×10^{-5}	−1.511	−1.856	−2.290

Table 1. Summary of optimum stripping voltammetry conditions for the determination of Pd(II), Pt(II) and Rh(III) with the constructed GC/Bi-AgFE bimetallic nanosensor [40].

3.4. Analytical features of the adsorptive stripping procedure

According to the literature, the determination of palladium by DPAdSV at the surface of the hanging mercury drop electrode (HMDE) was first described by Wang and Varughese [41]. Dimethylglyoxime was used as the complexing ligand in slightly acidic media (pH = 5.15) for the deposition of palladium-dimethylglyoxime complex (Pd-$(HDMG)_2$). In the present study, the determination of Pd-$(HDMG)_2$ was done in 0.2 M acetate buffer (pH = 4.7) solution at the surface of a bismuth-silver bimetallic nanosensor. The DPAdSV current of the Pd-$(HDMG)_2$ complex at optimal conditions yielded well-defined peaks, in the concentration range 0.4–1.0 ng L^{-1} shown in **Figure 3A**. The five concentrations used yielded a linear response and the equation of the linear calibration curve is $y = 0.773x + 0.6151$ with a correlation coefficient of 0.9911.

The differential pulse adsorptive stripping voltammetric (DPAdSV) current for the Pt-$(HDMG)_2$ complex was measured at optimal conditions using a bismuth-silver bimetallic nanosensor in **Figure 3B**. In these measurements, a series of Pt-$(HDMG)_2$ complex concentrations ranging from 0.2 to 0.8 ng L^{-1} in 0.2 M acetate buffer (pH = 4.7) solution with 30 s deposition time was used. The peaks observed in the differential pulse voltammograms are well defined and the five concentrations used yielded a linear response, and the equation of the linear calibration curve is $y = 0.690x + 0.718$ with a correlation coefficient of 0.9881.

The determination of the Rh-$(HDMG)_3$ complex in 0.2 M acetate buffer (pH = 4.7) solution was performed by DPAdSV analysis under optimized working conditions, and the voltammograms are shown in **Figure 3C**. Well-defined stripping peaks were observed at

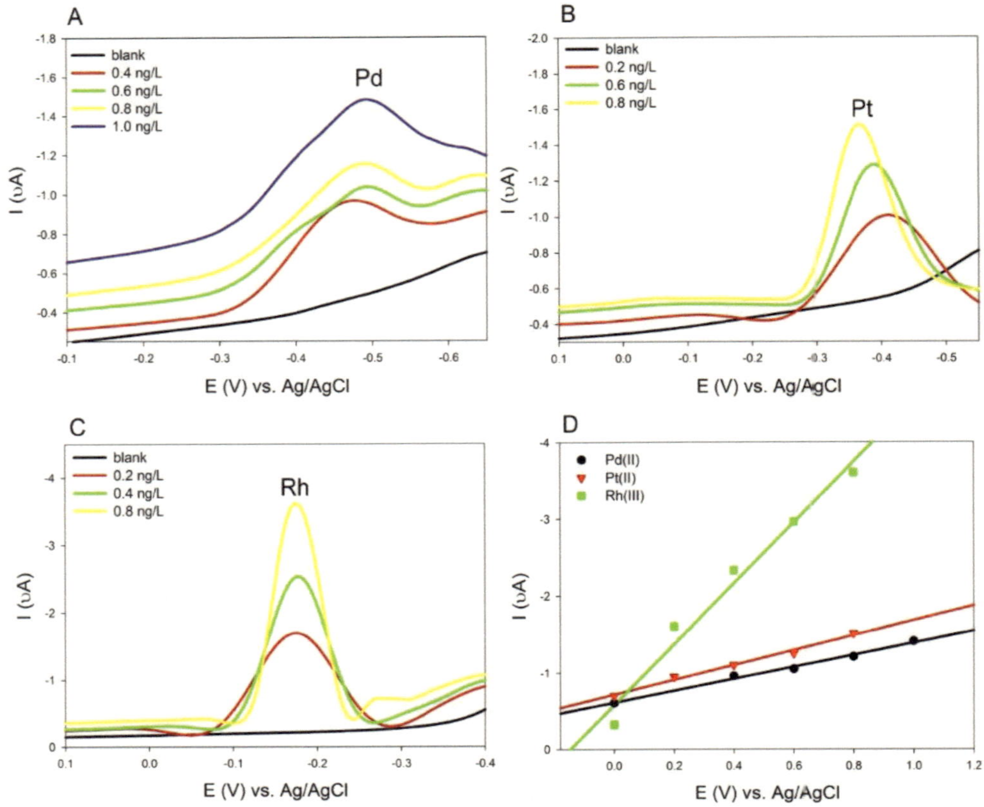

Figure 3. Differential pulse adsorptive stripping voltammetry results for increasing concentrations of (A) 0.4–1.0 ng L^{-1} Pd(II) with $E_d = -0.7$ V (vs. Ag/AgCl), (B) 0.2–0.8 ng L^{-1} Pt(II) with $E_d = -0.9$ V (vs. Ag/AgCl) and $t_d = 120$ s, (C) 0.2–0.8 ng L^{-1} Rh(III) with $E_d = -0.7$ V (vs. Ag/AgCl) and t_d 30 s at a SPCE/Bi-AgFE sensor, (D) corresponding calibration curves for the obtained DPAdSV curves. The electrolyte used consisted of 0.2 M acetate buffer (pH = 4.7) containing 1×10^{-5} M DMG concentration.

the bismuth-silver bimetallic nanosensor in the concentration ranging from 0.2 to 0.8 ng L^{-1}. The results indicated that dimethyglyoxime (DMG) can greatly promote the deposition of the Rh-$(HDMG)_3$ complex at the bismuth-silver bimetallic nanosensor and significantly increase the sensitivity of the determination of the Rh-$(HDMG)_3$ complex. The inset in **Figure 3C** showed that the DPAdSV peak currents have a linear response for the five concentrations evaluated, and the equation of the linear calibration curve is $y = 3.9527x + 0.5798$, with a correlation coefficient of 0.9703.

3.5. Interference and stability studies

In differential pulse adsorptive stripping voltammetry (DPAdSV), several trace metals can interfere with the determination of platinum group metals (PGMs) absorbing competitively onto the bismuth-silver bimetallic film electrode (Bi-AgFE) surface. They also complexing competitively with DMG producing signals close to that of the different PGMs or completely suppress the peaks. A number of metal ions that could potentially interfere with these PGMs were investigated such as Ni(II), Co(II), Fe(III), and Na+. The sulfates and phosphates were also investigated, and 1 ng L^{-1} of these interfering ions was added to the model solutions. These ions were chosen because they might reasonably be expected to exhibit redox activity at the SPCE/Bi-AgF sensor and exist in real samples. The behavior of Pd(II), Pt(II), and Rh(III) at concentrations of 0.5–1.5 ng L^{-1} in the presence of these cations and anions was investigated. This study showed that these ions have not interferes on the determination of Pd(II), Pt(II) and Rh(III).

The stability of the fabricated bismuth-silver bimetallic nanosensor was investigated for the peak current after every 7 h over a period of 28 h. The electrode was kept in deionized water after each measurement. Using the above optimized conditions, the bismuth-silver bimetallic nanosensor was utilized for the determination of 1 ng L^{-1} concentration of the PGMs evaluated. It was found that the peak current intensities decreased only slightly for the bismuth-silver bimetallic nanosensor, indicating that the nanosensor has good stability and repeatability (data not shown).

3.6. Analysis of environmental samples

The determination of PGMs was conducted in road dust and roadside soil samples collected in the Western Cape Province at Bottelary Road close to Stellenbosch and Old Paarl Road close to Klapmuts, outside Stellenbosch using the SPC/Bi-AgFE nanosensor. The bioavailability of the PGMs in the road dust and roadside soil samples was determined by subjecting the samples to a three-step sequential extraction procedure [42, 43].

The two sets of results obtained for road dust and roadside soil samples are shown in **Table 2**. The results for the dust and soil samples have shown that the method was successfully applied using the SPC/Bi-AgFE nanosensor. Relatively good results were obtained for

	DPAdSV		
	Carbonate bound	**Organic bound**	**Fe-Mn bound**
Sample	**Pd(II) (ng L^{-1})**	**Rh(III) (ng L^{-1})**	**Pt(II) (ng L^{-1})**
BOT1[#]	4.33 ± 0.21	4.39 ± 0.28	1.68 ± 0.07
BOT2[#]	3.68 ± 0.15	2.16 ± 0.27	0.90 ± 0.01
BOT3[#]	3.85 ± 0.20	12.68 ± 0.31	1.28 ± 0.01
BOT4[#]	4.25 ± 0.38	1.87 ± 0.38	0.56 ± 0.53
OP1[#]	2.56 ± 0.04	0.71 ± 0.10	1.54 ± 0.14
OP2[#]	3.05 ± 0.31	1.58 ± 0.08	2.39 ± 0.05
OP3[#]	2.74 ± 0.06	0.78 ± 0.10	8.78 ± 0.61
OP4[#]	4.29 ± 0.07	1.82 ± 0.32	2.96 ± 0.97

	DPAdSV		
	Carbonate bound	Organic bound	Fe-Mn bound
BOT1*	4.87 ± 0.22	5.98 ± 0.26	1.84 ± 0.12
BOT2*	3.68 ± 0.23	3.02 ± 0.29	1.18 ± 0.28
BOT3*	3.55 ± 0.42	15.93 ± 0.76	2.35 ± 0.26
BOT4*	4.55 ± 0.24	7.10 ± 0.45	1.23 ± 0.04
OP1*	2.82 ± 0.08	1.12 ± 0.07	2.17 ± 0.03
OP2*	3.87 ± 0.20	1.49 ± 0.15	2.57 ± 0.13
OP3*	3.22 ± 0.37	1.40 ± 0.14	5.09 ± 0.55
OP4*	4.53 ± 0.14	4.09 ± 0.54	1.08 ± 0.44

BOT, Bottelary Road; OP, Old Paarl Road.

Roadside dust.

* Roadside soil.

Table 2. Results obtained for the determination of PGMs concentrations using a SPC/Bi-AgFE nanosensor in dust and roadside soil samples collected from roads near Stellenbosch, Western Cape Province.

using the DPAdSV method. For high accuracy and sensitivity, only single PGM analysis was determined in all experiments using the constructed electrochemical sensor. These results indicate that the constructed SPC/Bi-AgFE nanosensor is more sensitive toward the determination of Pd(II), Pt(II) and Rh(III) in dust and soil samples.

3.7. Comparison of calculated results for different sensor platforms

To calculate the limit of detection (LOD), the formula 3σ/slope was employed, where σ is the standard deviation of the blank. The LODs of Pd(II), Pt(II), and Rh(III) obtained under the optimized conditions of these method were 0.7 ng L^{-1} for Pd(II), 0.06 ng L^{-1} for Pt(II), and 0.2 ng L^{-1} for Rh(III) for the SPC/Bi-AgFE nanosensor. In this study, the developed SPCE/Bi-AgFE nanosensor showed lower limit of detection than previously reported sensors based on the detection of PGMs in environmental samples. To illustrates the repeatability of the sensor, the relative standard deviation (RSD) was calculated and found to be 7.58% for Pd(II), 6.31% for Pt(II), and 5.37% for Rh(III) (n = 10). The reproducibility was evaluated using three different electrodes and a solution containing 1.0 ng L^{-1} of each metal ion with a RSD of 6.81% for Pd(II), 5.11% for Pt(II), and 5.97% for Rh(III).

The analytical performance of the SPC/Bi-AgFE nanosensor was compared with those obtained by other electrochemical sensors described in the literature for the determination of PGMs and illustrated in **Table 3** [35, 36, 44]. From the studies of modified electrodes, this SPC/Bi-AgFE nanosensor revealed lower limits of detection compared to the reported electrodes for PGMs determination. This developed procedure also reveals high sensitivity and faster response time for PGMs analysis in the presence of DMG, using acetate buffer (pH = 4.7) solution as the supporting electrolyte.

Electrode	Method	Linear range	LOD	References
SPC/Bi-AgFE	AdDPSV	Pd(II): 0.4–1.0 ng L^{-1} Pt(II): 0.2–0.8 ng L^{-1} Rh(III): 0.2–0.8 ng L^{-1}	Pd(II) – 0.07 ng L^{-1} Pt(II) – 0.06 ng L^{-1} Rh(III) – 0.2 ng L^{-1}	This work
GC/Bi-AgFE	AdDPSV	Pd(II): 0.2–1.0 ng L^{-1} Pt(II): 0.2–1.0 ng L^{-1} Rh(III): 0.4–1.0 ng L^{-1}	Pd(II) – 0.19 ng L^{-1} Pt(II) – 0.20 ng L^{-1} Rh(III) – 0.22 ng L^{-1}	[17]
SPC/BiFE	AdDPSV	Pd(II): 0–0.1 μg L^{-1} Pt(II): 0.2–0.1 μg L^{-1} Rh(III): 0–0.08 μg L^{-1}	Pd(II) – 0.008 μg L^{-1} Pt(II) – 0.006 μg L^{-1} Rh(III) – 0.005 μg L^{-1}	[35]
GC/BiFE	AdDPSV	Pd(II): 0–2.0 μg L^{-1} Pt(II): 0–3.5 μg L^{-1} Rh(III): 0–3.0 μg L^{-1}	Pd(II) – 0.19 μg L^{-1} Pt(II) – 0.20 μg L^{-1} Rh(III) – 0.22 μg L^{-1}	[36]
CGMDE	CAdSV	Pt(II): 2.0–80.0 μg L^{-1} Rh(III): 1.0–160.0 μg L^{-}	Pt(II) – 0.60 μg L^{-1} Rh(III) – 0.20 μg L	[44]
CGMDE	CAdSV	Pt(II): 5.0×10^{-12} to 1.5×10^{-9} mol L^{-1}	Pt(II) – 0.03 μg L^{-1}	[39]
Hg(Ag)FE	CSV	Pd(II): 1–50 μg L^{-1}	Pd(II) – 0.15 μg L^{-1}	[45]

Table 3. Comparison of results obtained in present work with other modified stripping voltammetric procedures for the determination of PGMs in model standard solutions and environmental samples are listed.

4. Conclusion

In conclusion, the construction, optimization, and practical application of the SPC/Bi-AgFE nanosensor, which were prepared by drop-coating onto a screen-printed carbon electrode, have been presented. The important DPAdSV parameters were optimized, and well-defined peaks were obtained for Pd(II), Pt(II) and Rh(III) in model standard solutions. To illustrate the practical application of the developed SPC/Bi-AgFE nanosensor, the sensor was tested for the detection of PGMs in road dust and roadside soil samples, collected from the Bottelary and Old Paarl Roads near Stellenbosch in the Western Cape Province. The results obtained for the developed nanosensor provide an alternative sensor platform to replace toxic mercury electrodes and can be used for routine determination of PGMs in road dust and roadside soil with high sensitivity.

Acknowledgements

The authors would like to thank the National Research Foundation (South Africa) and the CSIR (NRE, Stellenbosch) for supporting this research. We further acknowledge the support of the SensorLab, Chemistry Department, University of the Western Cape, Bellville, South Africa. A word of thanks also to the Chemistry Department, Faculty of Applied Sciences, Cape Peninsula University of Technology, Bellville, South Africa.

Author details

Charlton van der Horst[1], Bongiwe Silwana[2], Emmanuel Iwuoha[1] and Vernon S. Somerset[3*]

*Address all correspondence to: somersetv@cput.ac.za

1 SensorLab, Department of Chemistry, University of the Western Cape, Bellville, South Africa

2 Department of Chemistry, Durham University, Durham, United Kingdom

3 Department of Chemistry, Faculty of Applied Sciences, Cape Peninsula University of Technology, Bellville, South Africa

References

[1] A. Sonune, R. Ghate, Developments in wastewater treatment methods, Desalination, Vol. 167, 2004, pp. 55-63.

[2] S. Liu, L. Yuan, X. Yue, Z. Zheng, Z. Tang, Review paper. Recent advances in nanosensors for organophosphate pesticide detection, Advance Powder Technology, Vol. 19, 2008, pp. 419-441.

[3] A. Hildebrandt, R. Bragos, S. Lacorte, J.L. Marty, Performance of a portable biosensor for the analysis of organophosphorus and carbamate insecticides in water and food, Sensors and Actuators B, Vol. 133, 2008, pp. 195-201.

[4] M.A. El Mhammedi, M. Achak, M. Bakasse, A. Chtaini, Electroanalytical method for determination of lead(II) in orange and apple using kaolin modified platinum electrode, Chemosphere, Vol. 76, 2009, pp. 1130-1134.

[5] D. Dragoe, N. Spataru, R. Kawasaki, A. Manivannan, T. Spataru, D.A. Tryk, A. Fujishima, Detection of trace levels of Pb^{2+} in tap water at boron-doped diamond electrodes with anodic stripping voltammetry, Electrochimica Acta, Vol. 51, 2006, pp. 2437-2441.

[6] P. Gonzalez, V.A. Cortınez, C.A. Fontan, Determination of nickel by anodic adsorptive stripping voltammetry with a cation exchanger-modified carbon paste electrode, Talanta, Vol. 58, 2002, pp. 679-690.

[7] G. Marino, M.F. Bergamini, M.F.S. Teixeira, E.T.G. Cavalheiro, Evaluation of a carbon paste electrode modified with organofunctionalized amorphous silica in the cadmium determination in a differential pulse anodic stripping voltammetric procedure, Talanta, Vol. 59, 2003, pp. 1021-1028.

[8] G.H. Hwang, W.K. Han, J.S. Park, S.G. Kang, Determination of trace metals by anodic stripping voltammetry using a bismuth-modified carbon nanotube electrode, Talanta, Vol. 76, 2008, pp. 301-308.

[9] P. Sonthalia, E. McGaw, Y. Show, G.M. Swain, Metal ion analysis in contaminated water samples using anodic stripping voltammetry and a nanocrystalline diamond thin-film electrode, Analytica Chimica Acta, vol. 522, 2004, pp. 35-44.

[10] M.A. El Mhammedi, M. Achak, A. Chtaini, $Ca_{10}(PO_4)_6(OH)_2$-modified carbon-paste electrode for the determination of trace lead(II) by square-wave voltammetry, Journal of Hazardous Material, Vol. 161, 2009, pp. 55-61.

[11] Prior, C.E. Lenehan, G.S. Walker, Enhanced resolution of copper and bismuth by addition of gallium in anodic stripping voltammetry with the bismuth film electrode, Electroanalysis, Vol. 18, 2006, pp. 2486-2489.

[12] M.A. Baldo, S. Daniele, Anodic stripping voltammetry at bismuth-coated and uncoated carbon microdik electrodes, Analytical Letters, Vol. 37, 2004, pp. 995-1011.

[13] G. Aragay, J. Pons, A. Merkoçi, Enhanced electrochemical detection of heavy metals at heated graphite nanoparticle-based screen-printed electrodes, Journal of Material Chemistry, Vol. 21, 2011, pp. 4326-4331.

[14] X. Huab, S. Dong, Metal nanomaterials and carbon nanotubes-synthesis, functionalization and potential applications towards electrochemistry, Journal of Material Chemistry, Vol. 18, 2008, pp. 1279-1295.

[15] L. Rahman, A. Shah, S.B. Khan, A.M. Asiri, H. Hussain, C. Han, R. Qureshi, M.N. Ashiq, M.A. Zia, M. Ishaq, H-B. Kraatz, Synthesis, characterization, and application of Au–Ag alloy nanoparticles for the sensing of an environmental toxin, pyrene, Journal of Applied Electrochemstry, Vol. 45, 2015, pp. 463-472.

[16] S.N. Mailu, T.T. Waryo, P.M. Ndangili, F.R. Ngece, A.A. Baleg, P.G. Baker, E.I. Iwuoha, Determination of anthracene on Ag-Au alloy nanoparticles/overoxidized-polypyrrole composite modified glassy carbon electrodes, Sensors, Vol. 10, 2010, pp. 9449-9465.

[17] Van der Horst, B. Silwana, E. Iwuoha, V. Somerset, Bismuth-silver bimetallic nanosensor application for voltammetric analysis of dust and soil samples, Journal of Electroanalytical Chemistry, Vol. 752, 2015, pp. 1-11.

[18] N. Moghimi, M. Mohapatra, K.T. Leung, Bimetallic nanoparticles for arsenic detection, Analytcal Chemistry, Vol. 87, 2015, pp. 5546-5552.

[19] G. Darabdhara, B. Sharma, M.R. Das, R. Boukherroub, S. Szunerits, Cu-Ag bimetallic nanoparticles on reduced graphene oxidenanosheets as peroxidase mimic for glucose and ascorbic acid detection, Sensors and Actuators *B*, Vol. 238, 2017, pp. 842-851.

[20] C.Y. Yang, S-M. Chen, S. Palanisamy, B. Thirumalraj, X. Liu, Electrochemical synthesis of PtAu bimetallic nanoparticles on multiwalled carbon nanotubes and application for amperometric determination of nitrite, International Journal of Electrochemical Sciences, Vol. 11, 2016, pp. 4027-4036.

[21] L. Rahman, A. Shah, S.K. Lunsford, C. Han, M.N. Nadagouda, E. Sahle-Demessie, R. Qureshi, M.S. Khan, H-B. Kraatz, D.D. Dionysiou, Monitoring of 2-butanone using a Ag–Cu bimetallic alloy nanoscale electrochemical sensor, RSC Advances, Vol. 5, 2015, pp. 44427-44434.

[22] L. Daneshvar, G.H. Rounaghi, Z. Es'haghi, M. Chamsaz, S. Tarahomi, Fabrication a new modified electrochemical sensor based on Au–Pd bimetallic nanoparticle decorated graphene for citalopram determination. Materials Science and Engineering *C*, Vol. 69, 2016, pp. 653-660.

[23] Van der Horst, B. Silwana, E. Iwuoha, V. Somerset, Synthesis and characterisation of bismuth-silver bimetallic nanoparticles for electrochemical sensor applications, Analytical Letters, Vol. 48, 2015, pp. 1311-1332.

[24] C. Van der Horst, B. Silwana, E. Iwuoha, V. Somerset, Application of a Bismuth-silver nanosensor for the simultaneous determination of Pt-Rh and Pd-Rh complexes, Journal of Nano Research, Vol. 44, 2016, pp. 126-133.

[25] C. Van der Horst, B. Silwana, E. Iwuoha, E. Gil, V. Somerset, Improved detection of ascorbic acid with a bismuth-silver nanosensor, Food Analytical Methods, Vol. 9, 2016, pp. 2560-2566.

[26] V.S. Somerset, L.H. Hernandez, E.I. Iwuoha, Stripping voltammetric measurement of trace metal ions using screen-printed carbon and modified carbon paste electrodes on river water from the Eerste-Kuils River System, Journal of Environmental Science and Health Part A, Vol. 46, 2011, pp. 17-32.

[27] B. Silwana, C. Van der Horst, E. Iwuoha, V. Somerset, Inhibitive determination of metal ions using a horseradish peroxidase amperometric biosensor, in: T. Rinken (Ed.), State of the Art in Biosensors-Environmental and Medical Application, Intech, Croatia, 2013, pp. 105-119.

[28] B. Silwana, C. Van der Horst, E. Iwuoha, V. Somerset, Amperometric determination of cadmium, lead, and mercury metal ions using a novel polymer immobilised horseradish peroxidase biosensor system, Journal of Environmental Science and Health, Part A, Vol. 49, 2014, pp. 1501-1511.

[29] V. Somerset, C. Van der Horst, B. Silwana, E. Iwuoha, Biomonitoring and evaluation of metal concentrations in sediment and crab samples from the North-West Province of South Africa, Water Air and Soil Pollution, Vol. 226, 2015, p. 43.

[30] M. Noroozifar, M. Khorasani-Motlagh, A. Taheri, Determination of cyanide in wastewaters using modified glassy carbon electrode with immobilized silver hexacyanoferrate nanoparticles on multiwall carbon nanotube, Journal of Hazardous Materials, Vol. 185, 2011, pp. 255-261.

[31] F. Cui, X. Zhang, Electrochemical sensor for epinephrine based on a glassy carbon electrode modified with graphene/gold nanocomposites, Journal of Electroanalytical Chemistry, Vol. 669, 2012, pp. 35-41.

[32] S. Prakash, T. Chakrabarty, A.K. Singh, V.K. Shahi, Silver nanoparticles built-in chitosan modified glassy carbon electrode for anodic stripping analysis of As(III) and its removal from water, Electrochimica Acta, Vol. 72, 2012, pp. 157-164.

[33] M. Morfobos, A. Economou, A. Voulgaropoulos, Simultaneous determination of nickel(II) and cobalt(II) by square wave adsorptive stripping voltammetry on a rotating-disc bismuth-film electrode, Analytica Chimica Acta, Vol. 519, 2004, pp. 57-64.

[34] C. Locatelli, Simultaneous square-wave stripping voltammetric determination of platinum group metals (PGMs) and lead at trace and ultra-trace concentration level application to surface water, Analytica Chimica Acta, Vol. 557, 2006, pp. 70-77.

[35] B. Silwana, C. Van der Horst, E. Iwuoha, V. Somerset, Screen-printed electrodes modified with a bismuth film for stripping voltammetric analysis of platinum group metals in environmental samples, Electrochimica Acta, Vol. 128, 2014, pp. 119-127.

[36] C. Van der Horst, B. Silwana, E. Iwuoha, V. Somerset, Stripping voltammetric determination of palladium, platinum and rhodium in South African water resources, Journal of Environmental Science and Health Part A, Vol. 47, 2012, pp. 2084-2093.

[37] M. Georgieva, B. Pihlar, Determination of palladium by adsorptive stripping voltammetry, Fresenius Journal of Analytical Chemistry, Vol. 357, 1997, pp. 874-880.

[38] B. Ntsendwana, B.B. Mamba, S. Sampath, O.A. Arotiba, Electrochemical detection of Bisphenol A using graphene-modified glassy carbon electrode, International Journal of Electrochemical Science, Vol. 7, 2012, pp. 3501-3512.

[39] S. Huszal, J. Kowalska, M. Krzeminska, J. Golimowski, Determination of platinum with thiosemicarbazide by catalytic adsorptive stripping voltammetry (AdSV), Electroanalysis, Vol. 17, 2005, pp. 299-304.

[40] V. Somerset, B. Silwana, C. Van der Horst, E. Iwuoha, Construction and evaluation of a carbon paste electrode modified with polyaniline-co-poly(dithiodianiline) for enhanced stripping voltammetric determination of metals ions, in: K. Kalcher, R. Metelka, I. Švancara, K. Vytřas (Eds.), Sensing Electroanalysis, University Press Centre, Pardubice, Czech Republic, 2013/2014, pp. 143-154.

[41] J. Wang, K. Varughese, Determination of traces of palladium by adsorptive stripping voltammetry of the dimethylglyoxime complex, Analytica Chimica Acta, Vol. 199, 1987, pp. 185-189.

[42] L. Li, Z. Xu, J. Wu, G. Tian, Bioaccumulation of heavy metals in the earthworm Eiseniafetida in relation to bioavailable metal concentrations in pig manure, Bioresource Technology, Vol. 101, 2010, pp. 3430-3436.

[43] M.T. Morera, J.C. Echeverria, C. Mazkiaran, J.J. Garrido, Isotherms and sequential extraction procedures for evaluating sorption and distribution of heavy metals in soils, Environmental Pollution, Vol. 113, 2001, pp. 135-144.

[44] S. Huszal, J. Kowalska, M. Sadowska, J. Golimowski, Simultaneous determination of platinum and rhodium with hydroxylamine and acetone oxime by catalytic adsorptive stripping voltammetry (CAdSV), Electroanalysis, Vol. 17, 2005, pp. 1841-1846.

[45] A. Bobrowski, M. Gawlicki, P. Kapturski, V. Mirceski, F. Spasovski, J. Zarebski, The silver amalgam film electrode in adsorptive stripping voltammetric determination of palladium (II) as its dimethyldioxime complex, Electroanalysis, Vol. 21, 2009, pp. 36-40.